NATO ASI Series

Advanced Science Institutes Series

A series presenting the results of activities sponsored by the NATO Science Committee, which aims at the dissemination of advanced scientific and technological knowledge, with a view to strengthening links between scientific communities.

The Series is published by an international board of publishers in conjunction with the NATO Scientific Affairs Division

A Life Sciences B Physics	Plenum Publishing Corporation London and New York
C Mathematical and Physical Sciences D Behavioural and Social Sciences E Applied Sciences	Kluwer Academic Publishers Dordrecht, Boston and London
F Computer and Systems Sciences G Ecological Sciences H Cell Biology I Global Environmental Change	Springer-Verlag Berlin Heidelberg New York London Paris Tokyo Hong Kong Barcelona Budapest

PARTNERSHIP SUB-SERIES

1. Disarmament Technologies	Kluwer Academic Publishers
2. Environment	Springer-Verlag
3. High Technology	Kluwer Academic Publishers
4. Science and Technology Policy	Kluwer Academic Publishers
5. Computer Networking	Kluwer Academic Publishers

The Partnership Sub-Series incorporates activities undertaken in collaboration with NATO's Cooperation Partners, the countries of the CIS and Central and Eastern Europe, in Priority Areas of concern to those countries.

NATO-PCO DATABASE

The electronic index to the NATO ASI Series provides full bibliographical references (with keywords and/or abstracts) to about 50000 contributions from international scientists published in all sections of the NATO ASI Series. Access to the NATO-PCO DATABASE compiled by the NATO Publication Coordination Office is possible in two ways:

- via online FILE 128 (NATO-PCO DATABASE) hosted by ESRIN, Via Galileo Galilei, I-00044 Frascati, Italy.
- via CD-ROM "NATO Science & Technology Disk" with user-friendly retrieval software in English, French and German (© WTV GmbH and DATAWARE Technologies Inc. 1992).

The CD-ROM can be ordered through any member of the Board of Publishers or through NATO-PCO, Overijse, Belgium.

Series F: Computer and Systems Sciences, Vol. 148

The NATO ASI Series F Special Programme on
ADVANCED EDUCATIONAL TECHNOLOGY

This book contains the proceedings of a NATO Advanced Research Workshop held within the activities of the NATO Special Programme on Advanced Educational Technology, running from 1988 to 1993 under the auspices of the NATO Science Committee. The books published so far in the Special Programme are listed briefly, as well as in detail together with other volumes in NATO ASI Series F, at the end of this volume.

Springer
Berlin
Heidelberg
New York
Barcelona
Budapest
Hong Kong
London
Milan
Paris
Santa Clara
Singapore
Tokyo

Knowledge Acquisition, Organization, and Use in Biology

Edited by

Kathleen M. Fisher

Center for Research in Mathematics and Science Education
San Diego State University, 6475 Alvarado Road, Suite 206
San Diego, CA 92120-5006, USA

Michael R. Kibby

Centre for Academic Practice, University of Strathclyde
The Graham Hills Building, 50 George Street
Glasgow G1 1QE, Scotland, UK

Springer
Published in cooperation with NATO Scientific Affairs Division

Proceedings of the NATO Advanced Research Workshop on Biology Knowledge: Its Acquisition, Organization, and Use, held in Glasgow, Scotland, June 14–18, 1992

Library of Congress Cataloging-in-Publication Data

Knowledge acquisition, organization, and use in biology / edited by Kathleen M. Fisher, Michael R. Kibby.
p. cm. -- (NATO ASI series. Series F, Computer and system sciences ; vol. 148)
"Proceedings of the NATO Advanced Research Workshop on Biology Knowledge: Its Acquisition, Organization, and Use, held in Glasgow, Scotland, June 14-18, 1992"--T.p. verso.
"Published in cooperation with NATO Scientific Affairs Division."
Includes bibliographical references and index.
ISBN 3-540-60415-4 (alk. paper)
1. Biology--Study and teaching--Congresses. 2. Constructivism (Education)--Congresses. I. Fisher, Kathleen M. II. Kibby, Michael. III. North Atlantic Treaty Organization. Scientific Affairs Division. IV. NATO Advanced Research Workshop on Biology Knowledge: Its Acquisition, Organization, and Use (1992 : Glasgow, Scotland) V. Series: NATO ASI series. Series F, Computer and systems sciences ; no. 148.
QH315.K55 1995
574'.071--dc20 95-39014
CIP

CR Subject Classification (1991): K.3, J.3

ISBN 3-540-60415-4 Springer-Verlag Berlin Heidelberg New York

Printed in Germany

Typesetting: Camera-ready by editors
Printed on acid-free paper
SPIN: 10486135 45/3142 - 5 4 3 2 1 0

Preface

Biology education, like science education in general, is in the midst of a revolution that is worldwide in scope. The changes occurring in the ways in which science education researchers think about learning and understanding seem to many to be exciting and profound. They represent nothing less than a major paradigm shift, similar at least in some ways to that wrought by the molecular revolution in biology. In both fields technology led the way into a whole new level of understanding. The molecular biology revolution grew out of (and provided a channel for) increasingly sophisticated means of detecting, visualizing, and manipulating subcellular structures and complex organic molecules. Likewise, the learning revolution is to a large extent a consequence of our increasingly sophisticated mechanisms for representing, simulating, assessing, and interacting with cognitive processes and structures. As Francis Crick once said, the revolution allows us to ask questions at a level that was previously unimaginable.

As with most revolutions, it is much easier to see where we have been than where we may be going. The prevailing new theories reside largely under the umbrella of *constructivism*. Many science education researchers today agree with certain basic premises of constructivism, such as:

- cognitive knowledge must be created by the learner, as opposed to being passively received,
- new ideas are understood through their connections to prior knowledge,
- knowledge construction is effortful,
- knowledge construction is enhanced by practice, introspection, explicit analysis, and metacognitive skills,
- new ideas are best created in relevant contexts,
- our understanding of the world is informed by the accumulated artifacts of our cultures and in this sense cognition is distributed,
- social interactions form an important mechanism in the knowledge construction process,
- our model of the world can never be more than an approximation of reality (that is, there is no 'truth' in science), and
- useful knowledge is organized and can be readily applied in particular situations, as opposed to the inert knowledge promoted by texts and teaching methods of the recent past.

One of the problems researchers face is that the term *constructivism* is used to describe theories in at least three interrelated but nonetheless distinct domains. First, constructivism is a *theory of learning* embodying many ideas from cognitive, motivation/affect, and social theory. This is the brand of constructivism that constitutes the major focus of this book. Second, constructivism is a *theory of knowledge* – that is, an epistemology. This strand of constructivism also receives some attention here. Third, constructivism describes a *philosophy* which in its extreme form is called radical constructivism. There is almost no discussion of philosophy in this volume. The one direct philosophical reference, by Honey, refers to an older, alternative theory. Interestingly, one thing that constructivism is *not* is a theory of teaching. Yet it has many implications for instruction, some of which are explored in the following chapters.

The papers in this book come from researchers working in many different countries. In many cases they didn't know one another or one another's work prior to meeting in Scotland. Yet there is an amazing (to us at least) degree of continuity and complementarity in the ideas and data they present. From Rome to Toronto, from Baton Rouge to Hamburg, from early elementary grades to college, biology education researchers are devising ways to get students actively engaged in such activities as:

- explicating personal knowledge,
- struggling to make sense of phenomena, structures, and events (i.e., developing one's own situation models),
- asking questions (what do I need to know to understand…?),
- engaging in collaborative problem-based research to find and evaluate answers,
- developing skeletal representations and organizational frameworks of their emerging ideas, and
- applying ideas in new contexts.

Researchers consistently involve students in building upon what they already know, with the aim of developing relatively deep conceptual understanding of biological processes and events.

The reader will find the following chapters variable in terms of style and perspective. This is a collection of papers, not a continuous story. Authors include physics educators and psychologists as well as biology educators. Some chapters present research data while others are more theoretical. Yet each chapter presents a piece of the puzzle of how to produce deep conceptual understanding of biology. We can only hope the reader finds some of this information useful and enjoys the challenge of the puzzle.

The major themes of the book are summarized in Figure 1. The organization of chapters within these themes is shown on page VIII. This organization is somewhat arbitrary since many chapters could be placed in more than one category, but some organization is required and this is ours.

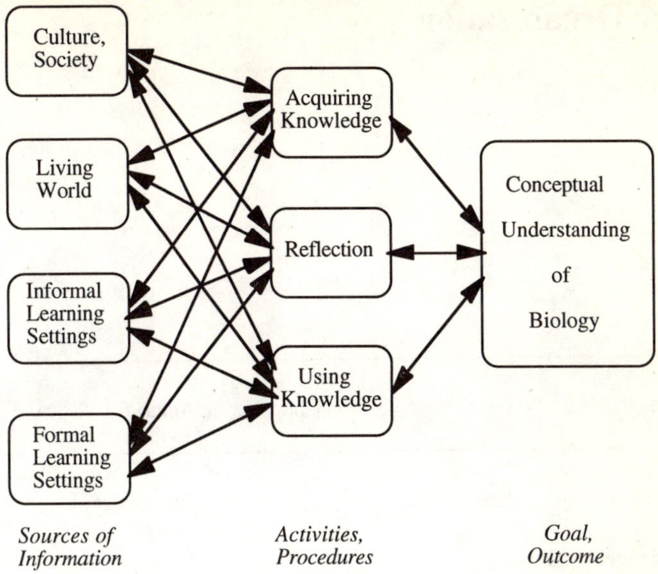

Figure 1. Facets of knowledge construction in biology

We think we can speak for all participants in expressing our enjoyment of the conference in Scotland. We are grateful to NATO for making the sharing of this work possible, and to the University of Strathclyde in Glasgow for the impeccable accommodations so conducive to open discussions. We are also grateful to the Center for Research in Mathematics and Science Education for their unwavering support in bringing this collection to press.

November 1995

Kathleen M. Fisher
Michael R. Kibby

Chapter Organization

Theme	Chapter	Author(s)
I. Overview of Biology Knowledge	1.	Jimenez et al.
II. Knowledge Acquisition		
From images	2.	Wandersee
From text	3.	Otero
From text, elementary / middle classes	4.	Scardemalia
In elementary / middle classes	5.	Jimenez
	6.	Brinkman
	7.	Arcà
	8.	Caravita
In secondary / college classes	9.	Buddingh
	10.	Gorodetsky & Fisher
III. Knowledge Organization		
Model of concept genesis	11.	Schaefer
Concepts and relations	12.	Brody
Nature of relations	13.	Faletti & Fisher
Conceptual graph structures	14.	Gordon
IV. Knowledge Use		
Mental models	15.	Honey
Retrospective causal reasoning	16.	Tamir

Table of Contents

The Nature of Knowledge in Biology and Its Implications for Teaching and Learning

Maria Pilar Jimenez Aleixandre[1], Ruth Amir[2], Michael J. Brody[3], Pinchas Tamir[2], and Warren Tomkiewicz[4]

[1] Department of Didactica das Ciencias Experimentals
Universidade de Santiago de Compostela, Spain
[2] Israel Science Teaching Center, The Hebrew University of Jerusalem, Israel
[3] College of Education, Montana State University, Bozeman, MT 59717, USA
[4] Department of Natural Sciences, Plymouth State College, USA

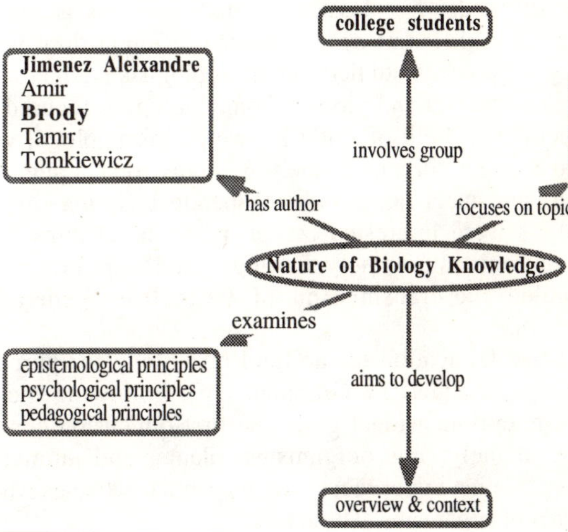

Abstract. This paper describes a number of epistemological, psychological and pedagogical principles considered critical to effective biology teaching. The authors illustrate the implementation of these principles in three case studies which focus on photosynthesis, evolution and estuarine biodiversity. Schwab's Four Commonplaces [46] are used to focus the discussion of how to improve teaching and learning of biological knowledge.

Keywords. Biology, Teaching, Learning, Epistemology, Cognitive psychology

The Life Sciences: Their Nature and Scope

Life sciences have been defined as those branches of science which deal with the study of living organisms and life processes, as well as their interactions with physical-inanimate objects and phenomena both in present and over time [58].

Perhaps the most interesting and astonishing feature of biology is the remarkable unity residing along with enormous diversity Unity is featured by the existence of a common pattern for all living organisms, exemplified by the cell as basic functional unit, in which the same mechanisms operate whether the cell is of a unicellular organism such as an amoebae or a huge animal such as an elephant, regardless of whether it belongs to a whale or to the minute plankton serving as food for this whale. Diversity is featured by the endless variety of forms, structures, colors, and behaviors exhibited by living organisms.

Historically, diversity was the focus of biological research which attempted to impose some structure and order by classifying the organisms according to certain dimensions. Typically, different people studied different dimensions or different fields of biology. Botanists studied plants, zoologists studied animals, and microbiologists studied bacteria and viruses. Some biologists focused on the study of forms and structures, a field of study known as Morphology, whereas others concentrated on tissues and organs, namely Anatomy; others attempted to understand the processes and mechanisms which operate in living organisms, namely Physiology; others were interested in the inter-relationships between organisms and their environment, namely Ecology; and still others in the continuity of life, studied the transmission of traits from generation to generation, namely Genetics.

A useful approach to the delineation of the field of biology for purposes of curriculum construction is presented by Grobman [25]. Following the BSCS approach, he presents a three-dimensional grid. The first dimension consists of the major three groups, namely: microorganisms, plants, and animals. The second dimension consists of nine major themes or conceptual schemes which cut across the different groups of organisms. These are:

(a) Nature of scientific inquiry
(b) The intellectual history of biological concepts
(c) Genetic continuity
(d) Regulation and homeostasis
(e) Complementarity of structure and function
(f) Behavior
(g) Relationship between organism and environment
(h) Diversity and unity
(i) Change through time – evolution

The third dimension, which cuts across the two already described, refers to seven hierarchical levels of organization from less inclusive to the more inclusive, which are: molecular, cellular, tissue and organ, individual organism, population, community, world biome. It is possible to specify objectives in terms of these three dimensions. It is also useful to analyze a given curriculum or textbook along these dimensions. For example, while conventional biology emphasizes organs, tissues ,and the individual organism, the BSCS Green version, the Australian Web of Life, and the Nuffield O-Level texts put great emphasis on populations and communities. At the same time, the BSCS Yellow version and the Nuffield A-Level put great emphasis on the cellular level and medium emphasis on the molecular level.

Special Features of Life Science in School

Every field of knowledge is distinguished by specific features which relate to the structure of the discipline and to its methods of inquiry. When that field of knowledge becomes a school subject, these features gain special flavor as they interact with the contexts provided by the school environment, thereby constituting the life science curriculum. A number of these features which characterize the life sciences as a curriculum will be discussed briefly below.

It is convenient to discuss the unique features of biology under the four commonplaces of education. Curriculum theories have long been known to rely on four fundamental conceptual reference points, called the *commonplaces of curricular thought*: (1) the student; (2) the subject or what is to be learned; (3) the milieu in which education takes place (the environment from the classroom to the entire society or culture); and (4) the teacher or other educative agent. Any set of ideas that gives meaning to guide and justify curricular practice must make some reference, explicitly or implicitly, to some conception of each of these commonplaces [64].

We shall present selected examples of the life sciences as a school subject classified by the dominant relevant commonplaces.

The Student Concept of Life

Life is the focus of biology education. Many teachers take it for granted that their pupils understand what life is. Yet it was found that while practically all children aged 9 to 15 classify animals as living, only 80 percent correctly classify plants as living and inanimate objects as nonliving, and still fewer (56 percent) classify embryos (seeds and eggs) as living. Movement and growth appear to be the most commonly used indicators of life. Only at the junior high level do students begin to employ as indicators of life more unique biological attributes such as nutrition, reproduction and respiration. For many children, the life of people and many animals is different from that of plants and embryos. Moreover, life as related to a dormant seed is conceived to be substantially different from that of the same seed as it germinates. The knowledge that living organisms develop from living organisms does not prevent many students from believing that seeds and eggs are not alive. There is no doubt that teachers should be aware of the ambiguities that children associate with the concept of life so that they will be able to choose more effective instructional strategies which will help students form a mature concept of life, which is essential for people's understanding of the phenomena around them [63].

Human Biology

Human biology as such has not been a common component of the biology programs of the 1960s. Rather it was assumed that human biology would best be studied as part of the general survey of the functional systems of living organisms. One reason for this is that it is often impossible to perform experiments with humans and since biology was to be studied to a large extent by laboratory investigation, it was considered reasonable to integrate human biology with that of other animals which lent themselves more easily to direct observations and experiments. As well, it was believed that a comparative study of human physiology, morphology, and anatomy would provide a useful evolutionary perspective, and perhaps help develop a sense of realism and humility in students who are often indoctrinated in other school subjects to adopt a highly homocentric view of the world. More recently, however, this approach has been widely criticized. Silverstein and Tamir [50] found that 43 percent of their sample of science educators, scientists, and teachers believed that human biology should be studied as a topic in its own right, while only 28 percent advocated integration with other biological topics. The desirability of a human biology course dealing with life processes as well as social, anthropological, economic, and cultural aspects of human life is voiced by many science educators [34, 50, 30].

Scientific Literacy

The uniqueness of the life sciences has already been attended to because of their immediate personal relevance to the students. Certainly biology as a major field of human knowledge is worth studying on its own merits. Some knowledge of major biological ideas and processes, some familiarity with the microorganisms, plants, and animals that surround us, and some insight into the structure and function of our own bodies are certainly (much like language, literature, and mathematics) part and parcel of human culture and of basic literacy. In addition, biology has the potential to make some unique contributions to education. Some of these potential contributions will be discussed briefly below.

While the physical sciences are by their nature less complex, the way they are being taught, their heavy reliance on abstractions, as well as their tendency to incorporate mathematics, make them unattractive to many students. It is well known that when they have a choice, many more students will elect to study biology in favor of either chemistry or physics. For example, in the United States about 90 percent of high school students study biology, about 50 percent study chemistry, and only about 20 percent study physics. This being the case, the study of biology carries the burden of not only presenting the content and processes of the life sciences, but rather of representing the only science that many students would ever experience in their high school studies. The image of science that many citizens carry into life would be the one acquired through their study of biology.

Students' Alternative Conceptions

A comparison of recent science educational journals (1986–1991) with journals in the previous 15 years (1971–1985) reveals areas which hardly existed in the earlier period, whereas in the more recent period they have flooded the literature. Among these areas have been 'children's science', 'naive conceptions', 'preconceptions', 'alternative frameworks', 'misconceptions', and other related titles. Hundreds of studies on alternative conceptions in the field of biology have been published [67]. The importance of these studies lies in their potential contribution to meaningful learning. Ausubel [6] emphasized how important it is for the teacher to ascertain 'what the student already knows' in order to be able to teach the learner accordingly; and thereby enhance meaningful learning by building on prior knowledge to tie the new material to existing 'anchor' concepts in the learner's cognitive structure. The present wave of research on students' conceptions concentrates on the interference, obstacles and constraints often exerted by existing concepts on the learning of related materials. Ideally, when students' preconceptions are known, the teacher can plan the instruction in such a way that any misconceptions which may exist are considered and treated so that the students will discard them as they are introduced to the alternative scientific concepts. If not adequately treated, the misconceptions will often make it impossible for the learner to understand and internalize the scientific concept.

The Subject Matter

History and Philosophy of Science

The main role of the history and philosophy of science is to direct students and teachers to understand the nature of science with special attention to the processes by which scientific knowledge is created. Notions, such as "the scientific method" or "proving that my hypothesis is correct" are incompatible with the nature of science, in which the diversity of research methods and the tentative nature of scientific knowledge claims is recognized. The philosophy of science deals with the nature and types of explanations, namely, causal, teleological, and anthropomorphic [60]. Finally, the history of science provides examples of the conceptions held by people in different periods, some of which are quite similar to children's ideas about and conceptions of scientific phenomena. Presenting such examples to students may help to persuade them to abandon misconceptions and be ready to adopt currently held scientific conceptions, realizing that these may also change in the future. As a side benefit, the acquisition of a realistic image of science increases students' confidence in their ability to study science, and improves their attitudes toward science by helping them to hold realistic expectations. Finally, for students who tend to shy away from science, stories about the life of a scientist may provide the "personal touch" which can raise their interest and readiness to study science in general and biology in particular, if the examples are taken from the life sciences. An example of how the history and philosophy of science can be incorporated into learning materials in biology may be found in Friedler and Tamir [23].

The World of 'Micro'

All physical and chemical processes have mechanisms which operate at the level of the micro', namely with tiny particles such as electrons, atoms, and molecules. Students cannot observe the interactions among these particles and therefore have to rely on inferences and abstractions. With the invention of microscopes, humans have gained the means to explore the level of the 'micro'. For most students this opportunity is provided within the framework of the life sciences, which offer ample opportunities to study the structures and functions of unicellular organisms as well as of cells and tissues in higher organisms. These studies, with the aid of the microscope, are not only important for the study of biology but serve also to extend the world view of students into the dimensions of the micro, very much like the extension of the world view of students beyond our own earth which is acquired through the study of astronomy. These two extensions are essential components of basic literacy needed by citizens in order to be able to function adequately in our world today.

The Learning Milieu

Outdoors – Laboratory – Classroom

Although other school subjects may also use a variety of learning environments, it is in the life sciences where the division of labor between classroom, laboratory, and outdoors has become accepted and desirable in many schools. Each of these environments offers unique learning opportunities, and only through a synthesis of these diverse faces of biology can the student acquire a faithful holistic image of biology.

The Use of Live Animals

The use of live animals poses a serious dilemma to biology teachers. Research has shown that most students are interested in studying live animals by direct observations and are highly motivated to experiment with animals, much more than with either plants or microorganisms [62]. Most students also believe that direct experiences involving the study of animals are superior to learning from secondary sources. At the same time, however, most students exhibit deep concern for and affection toward living organisms in general [59]. It follows that the use of live animals may have both positive and negative outcomes in terms of students' attitudes to life, to living animals, and to the study of biology [35]. Knowledge of the factors which affect students' attitudes toward the use of animals could help teachers in making more adequate decisions. Silverstein and Tamir [51] found, for example, that if the lead given by high school students is followed, animals should be used whenever their use appears helpful to learning, provided that undue suffering and cruelty are avoided and that precautions are taken to reduce the damage to animals as much as possible. The degree of usefulness to harmfulness associated with using an animal is the most important factor affecting students' attitudes to the use of that animal in observations and experiments.

Increased Emphasis on Science-Technology-Society (STS) Interactions

There are many good reasons for adopting a STS approach. Some of these are presented in the following citation: "Progress in science and technology (genetic engineering and nuclear weaponry, for example) forces us to confront rapid changes and raises what were once intellectual abstractions to the level of hard, often painful reality for individuals, families and nations. We often must make decisions about new knowledge and technologies that we have barely begun to understand, must less embrace" [37]. The major characteristics of an STS approach are:

- utilizing student-identified problems with local interest and scientific and technical components as organizers for the course;
- using local resources (human and material) as original sources of scientific or technical information that can be used in problem resolution;
- involving students in seeking scientific or technical information that can be used in problem resolution;
- involving students in seeking scientific or technical information that can be applied in dealing with real-life problems;
- extending science learning beyond the class period, the classroom, and the school;
- emphasizing career awareness, especially careers related to science and technology;
- providing opportunities for students to perform citizenship roles; and
- demonstrating that science and technology are major factors that will impact on the future.

The Teacher or Other Educative Agent

Knowledge of Content-Specific Pedagogy

The diversity of legitimate approaches to the teaching of biology makes it possible to adjust and adapt curricula and instruction to the needs and capabilities of any student at any age. The richness of the living world enables teachers to build upon the natural curiosity and intrinsic motivation that most pupils hold toward certain biological phenomena.

The biological sciences have special links to central human enterprises, such as medicine and agriculture, which may be considered in many ways as applied technologies of the life sciences.

Many cultural, economic, and moral issues of paramount importance to our everyday lives are deeply rooted and closely associated with biological processes. Taking advantage of these unique relationships may upgrade the lives of individuals and communities. Issues such as genetic engineering, nutrition, hygiene, and birth control, to mention a few, would often be best approached within the context of the biological studies.

Controversial Issues

The new biology, which begins to gain prominence in schools, also involves questions of ethics, values, morals, and aesthetics. While in previous years teachers felt responsible for disseminating biological knowledge in a way that created an image of science as an objective field of study, detached from the problems of everyday life, the new orientation opens the classroom for discussion and study of controversial issues such as evolution versus creationism, birth control, genetic engineering, chemical pest control, and the like. No discipline other than biology can provide the knowledge base needed for intelligent debates on these controversial issues.

Teaching by Inquiry

Although inquiry can be used in teaching any subject, its use in biology has been more common than in other subjects, probably due to the influence of Schwab [46]. Among other contributions, Schwab conceived the approach called 'invitations to inquiry'.

Invitations to inquiry are probably the most significant innovation in science teaching since the early 1950s. These are teaching units that bring before students small samples of the operations of biological inquiry, and which are adjusted to the competence and knowledge of the students, but at the same time provide open-ended situations which require students to be actively engaged in reasoning and problem solving. In completing the units, students make use of and at the same time systematically develop their inquiry skills. A detailed rationale and guidelines for their use, together with 44 planned instructional units, are presented in the *Biology Teacher Handbook* [46]. More recently the BSCS produced 40 single topic films and 20 inquiry slide sequences, each accompanied by a useful teacher guide. These audiovisuals are also constructed as invitations to inquiry. These unique instructional materials represent a real breakthrough in the attempt to lead science teachers to teach science as inquiry rather than dogma. These exemplary instructional materials may be one of the key reasons for the relatively higher success of curriculum reform in biology compared with that of the physical and social sciences.

The Use of Microcomputers

Microcomputers can be used in a variety of ways to enhance the learning of biology. Some uses, such as word processing or keeping track of students' progress, are common to all school subjects, There are, however, certain applications which are unique to science in general and to biology in particular. Following are some of these applications: (a) measuring variables such as temperature, humidity, and light by instruments hooked to a computer thereby obtaining results in tables and graphs; (b) using simulations and data bases especially in topics where processes of change take a lot of time such as genetics, feedback mechanisms such as neurobiology, and systems including populations and ecological systems; (c) designing tests to take advantage of the capability of computers to provide immediate feedback (tests of critical thinking and a combined practical computer test have been tried successfully in the teaching of biology in Israel); and (d) monitoring equipment and living material in the school laboratory (e.g. keeping schedules for periodic maintenance of equipment such as microscopes). So far very little use has actually been made of computers in biology teaching [49].

Certain topics, such as genetics and microbiology, do not lend themselves easily to laboratory work in school, mainly because of logistical difficulties. For example, cross-breeding may take weeks before obtaining results. In such cases, computer simulations may serve as an effective substitute.

Implications for Teaching

In the remainder of this chapter, several case studies of teaching particular biological topics will be described. In each case, an attempt will be made to relate to the four commonplaces and to apply some of the examples presented above to the specific issues of each case. The following topics will each serve as a focus of case studies: photosynthesis; evolution; and variation and estuarine biodiversity.

Case Study I: Photosynthesis

Photosynthesis is the most important biological process in the biosphere. Through photosynthesis energy from the sun is captured and transformed into chemical form which serves as the ultimate source of energy for all life processes. Most of the fuels used by modern human society are also products of photosynthesis in organisms that lived in the past. Unraveling the details of the marvelous feat that takes place inside the chloroplasts of green plants is an ongoing interest of the scientific community. It started with Van Helmont (around 1630) and Priestly (around 1771) and continues up until today. As recently as 1988 a Nobel prize was awarded to a team of researchers for their discoveries regarding the molecular structure of the photosynthetic reaction centers in bacteria.

The central importance of photosynthesis to life has made it an integral part of the biology curriculum, making it one of a few topics which is encountered by the students more than once during their school years. Some curricula introduce it as early as the fifth grade and it is included in most junior high as well as high school science curricula [54].

Including photosynthesis in the syllabus at different grade levels means that curriculum developers believe that, despite its complexity, photosynthesis can be taught at different levels of abstraction and teaching can be matched to the cognitive development and to the learning ability of the students.

Student Understanding

In the past 15 years a lot of research in science education was devoted to uncovering and identifying misconceptions among students. Most of this research centered around concepts in physics [42]. However, studies of biological concepts revealed similar problems pertaining to topics in biology such as: evolution [17, 16], genetics [56], the cell [18], osmosis and diffusion [24], and photosynthesis [5, 9, 7, 67].

The importance of photosynthesis and the fact that it is taught at different age levels are two reasons for the abundance of studies dealing with photosynthesis as compared to the other biological topics. A recent review by Wood-Robinson cites 22 articles dealing with problems in understanding plant nutrition and photosynthesis.

The survey of these and other articles reveals that two misconceptions were found in most studies and their incidence spans across varying ages and cultures. These are:

(1) Plants obtain organic materials ('food') from the soil.
(2) Photosynthesis is the respiration of plants.

The first of these is a naive idea that was often blamed for interfering with meaningful learning of photosynthysis [52]. The existence and resilience of this misconception should not be surprising. Firstly, from our daily life experiences we know that plants grow in soil, and their growth is promoted if we water them and add various (organic and inorganic) fertilizers to the soil. Moreover, this idea is not completely wrong. Plants do get water (a major component of vacuolated plant cells and a reactant in photosynthesis) and important inorganic substances from the soil.

Secondly, providing convincing evidence that products of photosynthesis are indeed the major source of plant materials is not a simple task. Two kinds of difficulties can be suggested. The first of these is a conceptual difficulty inherent in the process of photosynthesis. Carbon dioxide is an odorless, colorless gas which only exists in minute amounts (0.035%) in the atmosphere. How can one convince the student to accept the scientific conception that this gas is eventually converted into the materials of a large tree? For the student, the scientific conception of photosynthesis might seem much less plausible than the naive-alternative one and inconsistent with their prior experiences [43].

A second difficulty can be found in the school laboratory. Usually no long-term experiments are performed which show the gradual accumulation of materials as the plant grows. Reading and discussing the Van Helmont experiment that is part of many photosynthesis chapters is probably not enough to foster the necessary conceptual change [36].

The second misconception, that photosynthesis is the respiration of plants, was found in most studies starting at the junior high school level and above [6, 27, 22]. Again, one can think of several sources for this misconception. The first of these is the resemblance between the processes: Both respiration and photosynthesis have a gas exchange component in which the same gases, namely, oxygen and carbon dioxide, take part. The term 'breathing' is used for this component of respiration. For students who are not aware of the complex biochemical reactions of both processes, the fact that in photosynthesis CO_2 is taken in and in respiration CO_2 is exhaled is probably not enough to cause conflict. When students first study the biochemistry, however, they often develop conceptions which confuses both processes.

Among older students, variations of this misconception develop. Examples are: photosynthesis takes place only in the day and respiration takes place only at night; or plants respire only at night or only in the dark. What might be the source of this misconception? In this case, it can be hypothesized that the overall chemical equation of photosynthesis and respiration can lead to the idea that one process is 'the opposite' of the other and therefore both cannot operate simultaneously. In fact, phrases that imply that one process is the opposite of the other can still be found in textbooks [41]. Among students who have learned about 'light' and 'dark' reactions of photosynthesis we [1] have found the following misconception: The 'light' reaction is photosynthesis and the 'dark' reaction is respiration.

As is the case with the source of materials for plant growth, inappropriate learning opportunities might contribute to the prevalence and resilience of this confusion. In many school laboratory investigations, the rate of gas exchange is the only measure associated with the study of photosynthesis and respiration. Together with the crucial importance of oxygen to life, this leaves the students with the impression that gas exchange is the major attribute of both photosynthesis and respiration.

In our own research on problems pertaining to learning photosynthesis among high school students (ages 17–18) during the last two years of secondary education [1, 3], we found additional problems and misconceptions relating to the interrelationship of photosynthesis and transpiration and to the understanding of the concepts autotroph and 'limiting factor'.

It is quite reasonable to assume that the difficulties and misconceptions described above do not exist in isolation. The contrary might be the case: students who view photosynthesis as merely a gas exchange process might also hold the notion that plants obtain materials for plant growth from the soil. Moreover, they might not meaningfully understand the link between autotrophy and photosynthesis. Evidence for such connections between misconceptions was found in the above mentioned research [1].

Epistemology and Psychology

In the preceding section we have touched upon several possible sources of misconceptions of photosynthesis. We can now turn to the contribution of theories of knowledge construction and learning, to meaningful learning and conceptual change.

Many authors follow Ausubel's advice as well constructivist ideas and suggest that teaching and learning should start with a phase of uncovering 'what the student knows', namely his current conceptions and prior knowledge [19]. The importance of this phase relies on the following assumptions:

(a) Prior knowledge contributes to subsequent learning [6], and
(b) For conceptual change to occur, students must first become dissatisfied with their alternative/erroneous (mis)conceptions [43].

Awareness on the part of the teacher of misconceptions about photosynthesis can contribute to his/her ability to handle naive ideas. Tamir [58] has recently recommended that knowledge about misconceptions should be an integral part of the teacher subject-matter-specific pedagogical knowledge.

Quite some time ago it was found that students have a low preference for learning about plants as compared to animals or the human body [62]. This disposition poses an obstacle to learning about photosynthesis. In addition, Stavy et al. [55] have raised the possibility that young students are reluctant to accept that the highly developed humans are dependent on "simple" organisms such as plants.

Bringing about conceptual change and/or reversing attitudes is not an easy task. We would like to suggest that some of the problems summarized in this paper can be overcome by developing learning materials and learning opportunities which will treat photosynthesis not as an isolated and rather awkward plant process but from one of four angles. Each of the four approaches holds its own advantages from both psychological and epistemological points of view:

(1) *Interdisciplinarity.* Understanding photosynthesis requires a broad base of biological, chemical and physical knowledge. Food, energy, organic materials, light and gas are a few basic concepts that are needed as prerequisites [53]. For example, teaching about photosynthesis with "energy" as a central theme can utilize this knowledge base. Quite often energy is perceived as having different meanings in biology, chemistry and physics. By stressing the common meaning

of energy, independent of discipline, integrative reconciliation can be promoted [6]. This is an important part of the process of meaningful learning.

(2) *Levels of organization.* One characteristic of biology is the different levels of organization. Introducing photosynthesis as an ecological process that begins most food chains and drives the energy flow between communities might be the preferred path for young or less able students. Only after this role is established should details of the process be introduced. Choosing this approach might also form more stable links in the student's cognitive structure, between ecological concepts such as autotroph on one hand and photosynthesis as a biochemical process (inside the chloroplasts of the plants' cells) on the other hand.

According to this approach, learning begins with concrete organisms and events (e.g. growth, material accumulation and gas exchange), and as learning proceeds towards the lower levels of organization (e.g. cells and organelles) the subject matter becomes increasingly more abstract.

(3) *History of Science.* The gradual discovery of photosynthesis is described in detail in many textbooks [33]. In this respect photosynthesis is quite unique among biological topics. This historical chain of events, as summarized by Luken (1988), can be used to design a teaching sequence which parallels major discoveries: The gas exchange, the role of CO_2, the role of light and chlorophyll, the association between red and blue light and oxygen release, the source of the oxygen, the details of carbohydrate synthesis, and the association between light reactions and electron flow. By tracing the process from the most general and inclusive aspects to the most specific ones, the historical sequence can facilitate meaningful learning [6]. Employing a historical approach also provides opportunities for discussing with the students the science processes as well as the tentativeness and ever-changing nature of scientific knowledge.

(4) *Science, Technology and Society (STS).* Teaching science using an STS approach has gained momentum in recent years. Here the topic of photosynthesis can be the focus of issues such as: How can we provide food for the rapidly growing world population? What might be the effect of global warming, deforestation and of increased CO_2 concentrations on the biosphere? How can we conserve fossil fuels? One of the most appealing attributes of this approach is its relevance to the student's life. This, in turn, increases motivation and encourages the development of positive attitudes.

Case Study II: Variation and Evolution

What observations is the theory of evolution attempting to explain? While working on research on the learning of natural selection, we were struck by the fact that a great number of pupils and teachers found difficulties in answering such a question, and ended up saying something as 'the change of living beings' [32]. No wonder, because we find that textbooks rarely relate the theory of evolution to the question it tried to answer: What accounts for the diversity and variety of living organisms?

Biodiversity is now an important issue for us, as concern about the environment increases. In Darwin's time the prevailing explanation for why so many different species existed was fixism: the numerous species had been created at the beginning of time in the same form as they exist today. The transformist (later called evolutionary) explanations of Lamarck and Darwin were quite different. Transformists claim that organisms that exist today differ from their ancestors in the past, and further that similar organisms (biological relatives) have common ancestors. Living beings change as a result of chance and differential survival under different conditions, that is natural selection.

Thus, Darwinism can be seen as an attempt to explain the diversity of life, the origin of so many different plants and animals; this is in fact the title of Darwin's book *The Origin of Species*. This kind of variation among organisms is the interspecies variability that makes a cat different from a mouse, or a marine periwinkle different from a land snail.

A fundamental step toward understanding the mechanisms that gave rise to varying species over time and toward understanding the entire process of natural selection is to acknowledge the importance of the other form of variation, the one which exists among different individuals of a single species. Differences in such characteristics as color, size, tolerance for changes in temperature, and resistance to chemicals provide the genetic variability upon which natural selection and differential survival may operate. When there are no options to choose from, there is no differential survival, and therefore no change in the population. In fact, when a species loses its genetic variation, its ability to respond to changing environmental conditions and therefore its own survival is challenged. This is the case of the threatened cheetah, whose individuals share a genetic pool with too little variability, too few heterozygotic loci, within it.

In other words, there are two kinds of variation related to the understanding of evolution. The first is the variation between organisms belonging to different species, whose origin the theory of evolution attempts to explain. The second is the variation among members of each species, that constitutes the raw material for the operation of the mechanisms of natural selection.

Teaching Variation

In our opinion, to learn the mechanism of natural selection in a meaningful, *functional* way (and we call it functional when the knowledge can be applied to contexts different from the one employed in instruction) implies an understanding of variation not only between different species, but also and particularly among individuals of a single species. The following discussion is informed by research on the learning of evolution in a classroom study involving a teaching sequence that includes, as one of its main features, attention to pupils' own ideas [32].

To say that there is a strong relation between the teaching of evolution and the teaching of genetics is commonplace. But in many textbooks and syllabuses (especially in Spain), not only are the two topics taught separately, but the emphasis when discussing biological inheritance is on the conservation of similarities and not also on the fact that genetic variation is generated by the inheritance mechanisms. It would be interesting to emphasize that, just as there are mechanisms to ensure the faithful transmission of information from one generation to the next, there are other mechanisms, such as random distribution of chromosomes in the gametes and crossing over, that ensure differences among individuals.

In both topics, genetics and evolution, two of the difficulties pupils encounter are (a) reasoning in probabilistic terms, and (b) thinking in terms of populations. The instruction doesn't help to overcome these difficulties, because the examples we usually find in texts of genetics problems relate to the offspring of one pair, and the question discussed is how individuals present a certain trait; also the phenotypic proportions are presented as fixed numbers, in a rather deterministic way. Teachers know that the proportions of different phenotypes are probabilities we expect, and that evolutionary mechanisms operate at the population level, not at the level of one individual or one pair, but we must find a way to make this implicit knowledge explicit for pupils, to transform it from something we take for granted, to knowledge they are able to operate with.

A variety of approaches could be used to promote this meaningful learning. For instance, to introduce the evidence for intraspecies variation, we have employed such activities as:

- Counting the number of seeds in pea sheaths,
- Listing different edible varieties of a species like Brassica oleracea: green and red cabbages, cauliflower, broccoli, brussel sprouts, and
- Filling a worksheet with the way in which each pupil presents a number of inherited traits like eye and hair color, ear lobe, blood group, sex etc. to evidence that there is little chance of finding in a classroom two individuals with the same combination of only a dozen traits.

The emphasis on variation also helps students to understand that one trait cannot be called ‘advantageous’ or ‘disadvantageous’ in a vacuum, but that it would be meaningful to do so only in the context of a certain environment, or in comparing what may be advantageous in one environment and disadvantagous in another. We used problems and simulation games which help pupils to experience the various paths that a species can take under different conditions. Preparing laboratory experiences is exceedingly difficult for all topics related with evolution, but we found very helpful the use of thought experiments, and of such strategies as Retrospective-Causal Reasoning [61].

Of course, we shouldn't expect that secondary school pupils learn, that is construct their own meanings for the mechanisms of biological inheritance and biological change, without difficulties. These are complex topics, and it took a great deal of hard thinking and experimenting for scientists to build the models we use today. When designing instruction we should take into account these difficulties and involve pupils in activities that can help them to overcome the difficulties.

Case Study III: Estuarine Biodiversity

Biodiversity is a prevailing scientific conception which has great importance in the future of the planet. This has become very evident at recent international meetings focusing on the earth's ecology (Earth Summit 1992). Biodiversity is critical to understanding biology and occurs at all levels of biological organization from molecules to entire ecosystems [44].

Student Understanding

In the study of a complex idea such as pollution in biology and particularly ecology, consideration should be given to what a student knows as an integrated set, a conceptual framework, or a cluster of concepts related to the idea. That is, student knowledge should reflect the complex, integrated and multidisciplinary reality of such natural phenomena. In the process of completing several studies concerning children's conceptions of ecology, we have found that seldom do students see the multiple connections that such concepts have to the real world. Students are unable to explain higher order concepts related to biodiversity, such as food webs, nutrient cycling or productivity, in terms other than those provided in their text [13]. They are unable to apply their knowledge to everyday occurrences in their lives. Yet, when probed in interviews concerning the nature of local environmental issues, children have rich understandings of a number of concepts from a variety of disciplines [14].

Similar studies related to comprehensive, interdisciplinary assessment of children's conceptions have been conducted in the Learning in Science Project. If we tie together several typical misconception studies, for example, we find that students have limited conceptions of the nature of light and plants [9]. This can lead to further misconceptions about the distribution of estuarine plants as a function of light penetration at various water depths. Studies of children's conceptions recently reported in many popular journals have given a better understanding of what we can expect children to understand concerning specific science concepts [24, 28, 65].

Teaching Biodiversity

For one week in July 1992, the authors taught an inservice course for secondary science teachers in the state of New Hampshire, USA. During that week, lecture, laboratories and field work concentrated on the estuarine biodiversity of Great Bay, NH. Great Bay is the largest estuary in the New England region of the US and as such has great significance to the ecology of the area. Recently it has been designated as part of the National Estuarine Research Reserve system by the federal government. In addition, the US Fish and Wildlife Service has designated a large part of the shoreline of Great Bay as a National Wildlife Refuge. It is an area of multiple use including a commercial oil port facility, recreation including outstanding striped bass fishing, defense including a nuclear submarine facility, and education including an estuarine research laboratory and a new US Fish and Game Interpretive Center.

In order to help teachers in the inservice course better understand the complex issues surrounding biodiversity and the Great Bay, the authors used a variety of educational techniques to teach the course. Among these were the use of recent scientific research, extensive field work, laboratory models of the estuary, concept mapping [40] and Gowin's Vee [26].

The use of current, exemplary scientific research is critical when teaching about emerging conceptions such as biodiversity. In this case the most comprehensive scientific works on Great Bay were distributed to the teachers in the course several weeks before it began. Among those publications was the recently published *An Estuarine Profile and Bibliography of Great Bay, New Hampshire* [47]. This allowed the teachers to read and learn about the estuary before their arrival at the course. The readings were interdisciplinary and focused on modern conceptions of biodiversity and the estuarine system, such as the hydrochemistry, biogeochemical processes, hydrosystem, geomorphology, and estuarine habitats and organisms.

Concept maps (Figure 1) were used to analyze the conceptual aspects of the research. The authors constructed maps before the course began to better understand the complex relationships within the system they were going to teach. These maps were constructed using pencil and paper as well as computer based concept mapping strategies. The concept maps were used during the course to negotiate meaning of the material with the teacher/students in the course. In addition to helping share meaning, the maps were also used to teach basic concept mapping strategies.

Each day during the course the majority of time was spent in the field investigating the geological, physical, chemical, biological and ecological components of the system. Among the activities performed by the teachers were taking cores of marine sediments, performing physical and chemical water quality tests, and obtaining transects of various habitats. Field work comprised approximately 75% of each days activities. The field studies were integral for the class to study and eventually appreciate the complexity of studies focusing on biodiversity. Transects throughout the estuary are essential for developing an understanding of the ecology based on differences in structural and functional diversity over time and space [44]. The functional values of various parts of Great Bay are determined by the physical characteristics and biological structures found within those parts.

Laboratory work focused on the construction of a model of the estuarine system. The focus was on setting up of marine aquaria [15] based on the field investigations. Gravel sediments were collected for the biological filtering system, water was collected in the estuary and marine organisms from the transect studies were kept in the marine aquaria models. Models were then modified to replicate the ecological "box" that Great Bay exhibits. Analysis of changes over time within the model are then conducted.

Gowin's Vee analysis was used to show the epistemological components of scientific research and how it relates to field and laboratory work. Students in the course were presented with an analysis of recent estuarine research by the authors (see Figure 2 below). Our intention was to share meaning with the students over an important scientific study related to Great Bay as well as teach

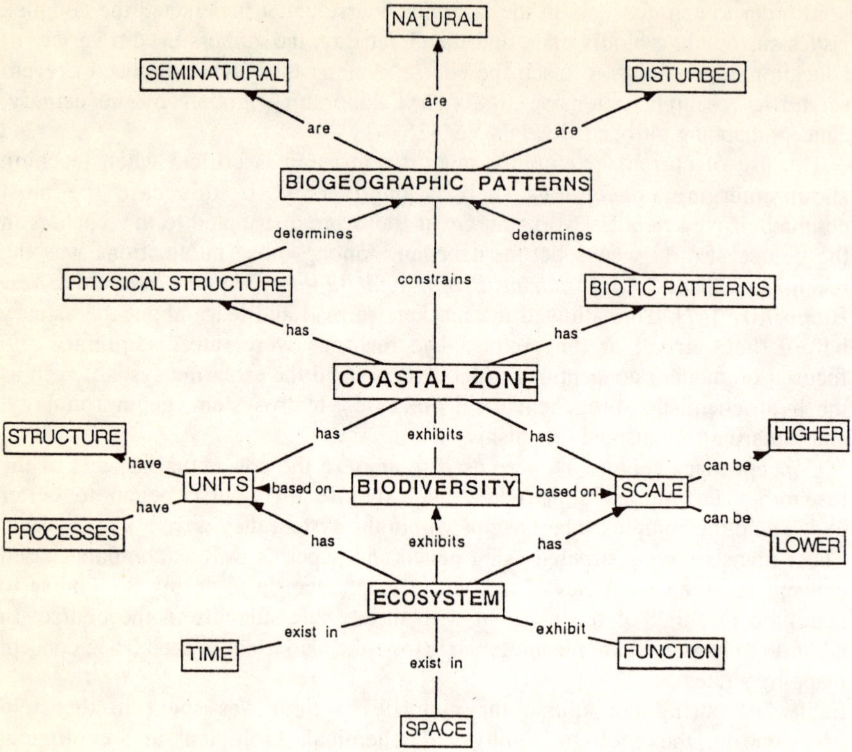

Figure 1. Concept map of major concepts related to estuarine biodiversity

TOPIC: Eelgrass Wasting Disease

REFERENCE: Short, F.K. et al., 1987, Eelgrass wasting disease: cause and recurrence of a marine epidemic, *Biol. Bull.*, **173**: 557-562.

CONCEPTUAL

PHILOSOPHY:
Marine Ecology

THEORY:
Population Dynamics
Estuarine Ecosystem

PRINCIPLES:
epidemiology
pathogenic organisms
temperature changes
reduced light
combined environmental factors
environmental stress
axenic culture
Koch's postulates
mass mortality

CONCEPTS:
pathogenic, nonpathogenic, microorganisms, infection, strain disease, healthy, Atlantic coast Pacific coast,Maine, New Hampshire, Massachusetts, Puget Sound migratory waterfowl, commercial fisheries scallops, nearshore soft sediments infection, leaves, mortality, necrotic lesions salinity, time, pollution, shading, habitats

Focus Questions

1. What causes masive mortality of eelgrasss populations on the North Atlantic coast?

2. What disease currently infects eelgrass populations and produces symptoms reminiscent of eelgrass wasting diseaseof the 1930s?

METHODOLOGICAL

VALUE CLAIMS:
Current wasting disease poses a major threat to coastal fisheries, waterfowl populations, numerous marine habitats, and the health of the estuarine environment.

KNOWLEDGE CLAIMS:
1. There is widespread infection of eelgrass with wasting disease.
2. There are both pathogenic and nonpathogenic strains of *Labyrinthula* widely distributed in the estuarine environment.

TRANSFORMATIONS

Labyrinthula source: substrate, location	#replic.	%infect.
diseased eelgrass, Great Bay, NH	33	100
diseased eelgrass, Puget Sound, WA	8	100
diseased eelgrass, Beaufort, NC	4	100
health eelgrass,Beaufort, NC	4	0
Spartina, Sapelo Island, GA	6	0
Codium drift, Weekapaug, RI	9	0
mangrove leaf, Longboat Key, FL	7	0

RECORDS:
1. number of replicates
2. percent infection
3. observation of lesions

EVENT
1. collection of specimens in field.
2. Laboratory infection of eelgrass, *Zostera marina* L., by *Labyrinthula.*

Figure 2. Vee analysis of a recent scientific study related to estuarine biodiversity

students about the use of Gowin's Vee (Figure 2) in relating scientific research with field and laboratory work.

To reiterate, this case study exemplifies many of the strategies previously identified as critical to improving biology education. Among these are the modern conceptions of biology such as biodiversity, the relevance of the material for society, the role of conceptual and methodological aspects of the scientific endeavor, focus on the major themes and levels of organization in biology and finally the use of educational strategies which allow for the construction of meaning by the students.

Conclusion

These three studies exemplify to some degree those principles which the authors consider to be critical to improving biology education, based on recent advances in epistemology and cognitive psychology. All three case studies represent those branches of science which deal with living things and their interrelationships, and take into consideration the unique nature of biological inquiry.

In this analysis children's conceptions of life, life processes and ecological interrelationships are critical for promoting meaningful learning. Several attempts have been made to overcome misconceptions and to improve understanding of photosynthesis [45, 10, 11, 20] with varying degrees of success [8, 22, 3]. Children's understanding of ecology has been the focus of several studies which indicate little understanding of complex interrelationships and application of basic concepts to the real world [12]. Research about misconceptions, together with learning theories and conceptual change models provide a lot of working ground for further improvement of teaching of this vital and complex process.

The nature of biological knowledge is complex and must be addressed directly by teachers and learners. Biology and particularly ecology represent complex phenomena which require an understanding of both a variety of natural sciences as well as their interactions. In teaching biology, teachers must identify essential concepts which are important to the understanding of each topic. Although at first glance each topic may appear distinct from others, through the course of analyzing related scientific research like that involving biodiversity, the construction of concept maps and Vee diagrams, it is apparent that there are a number of essential concepts that are common to a number of topics. This often suggests conceptual reorganization of both content and pedagogical knowledge, a process that effective teachers continue to practice throughout their careers. As this process evolves and their understanding becomes more complex, the use of paper and pencil concept mapping and Gowin's Vee diagram techniques may be limiting, and the use of new computer-based technologies, such as the SemNet software, which represent knowledge structures in an unlimited number of dimensions become essential.

The school context of teaching biology must include a variety of learning environments including the outdoors and laboratory. Collection and maintenance of organisms is essential for modeling biological and ecological phenomena.

Since science is complex, instruction must seek to provide opportunities for hands-on experiences with real organisms. Modeling of life processes and their interrelationships is essential for their understanding.

Finally, biology and ecology and their relationship to the real world are an increasingly important topic for science teachers to consider in light of the future of civilization and the well being of the planet. Given the prevalence in mass media of scientific concepts related to everyday events, it is imperative to have societies which understand the role of science, and particularly biology, in the real world.

References

1. Amir, R.: Identifying problems pertaining to learning and understanding of the concept 'photosynthesis' and developing ways to deal with them in the classroom. Ph.D. Thesis. Submitted to the Senate of the Hebrew University, 1991.
2. Amir, R., Tamir, P.: When does a factor become a 'limiting factor'? A study of students' misconceptions. Journal of Biological Education, 23: 129-134, 1989.
3. Amir, R., Tamir, P.: Development and evaluation of research based materials for teaching photosynthesis to senior high school students. Paper presented at NARST, Cambridge MA, March 1992.
4. Anderson, C.W., Sheldon, T.H., DuBay, J.: The effects of instruction on college non-majors' conceptions on respiration and photosynthesis. Journal of Research in Science Teaching, 27: 777-787, 1990.
5. Arnold, B., Simpson, M.: An investigation of the development of the concept photosynthesis to S.C.E. 'O' grade. Aberdeen College of Education. Aberdeen, Scotland, 1980.
6. Ausubel, D.P.: Educational Psychology: A Cognitive View. New York: Holt, Rinehart & Winston 1968.
7. Barker, M.: Teaching and learning about photosynthesis. Working papers no. 220-229. Science Education Research Unit. University of Waikato, Private Bay, Hamilton, New Zealand, 1984.
8. Barker, M., Carr, M.: Teaching and learning about photosynthesis. Part 2: A generative learning strategy. International Journal of Science Education, 11: 141-152, 1989b.
9. Bell, B., Brook, A.: Aspects of Secondary students' understanding of plant nutrition: Full Report. Centre for Studies in Science and Mathematics Education. The University of Leeds, 1984.
10. Bishop, B.A., Roth, K.J., Anderson, C.W.: Respiration and Photosynthesis. A teaching module. Institute for Research on Teaching. Michigan State University, 1985.
11. Boschuizen, R., Harms, A.H., Gier, A.H., Brinkman, F.G.: The influence of a series of lessons reckoning with preinstructional ideas on the development of the concept of plant nutrition. Paper presented at ATEE, Limmerick, Ireland, 1990.
12. Brody, Michael: Student Science Knowledge Related to Ecological Crises, International Journal of Science Education (in press).

13. Brody, M.: A programmatic approach to understanding student knowledge of environmental issues, Proceedings of the Second International Conference on Misconceptions of Science and Mathematics, Cornell University, Ithaca, NY, 1987.
14. Brody, M., Marion S., Chipman E.: Student knowledge in fourth, eighth and eleventh grades related to acidic deposition, Journal of Environmental Education, 20 (2), Winter (1988-89).
15. Brody, M., Paterson, B.: Aquarium Aquaculture, The Science Teacher, 36-41, 1992.
16. Brumby, M.: Problems in learning the concept of natural selection. Journal of Biological Education, 13: 119-122, 1979.
17. Deadman, J.A., Kelly, P.J.: What do secondary school boys understand about evolution and heredity before they are taught the topics. Journal of Biological Education, 12:7-15, 1978.
18. Dreyfus, A., Jungwirth, E.: The pupil and the living cell: A taxonomy of dysfunctional ideas. In: Proceedings of the Second International Seminar: Misconceptions and Educational Strategies in Science and Mathematics (Novak, ed.), Vol. 2, pp. 91-96. Ithaca, NY: Cornell University 1987.
19. Driver, R., Oldham, V.: A constructivist approach to curriculum development in science. Studies in Science Education, 13:105-122, 1986.
20. Eisen, Y., Stavy, R.: Students understanding of photosynthesis. The American Biology Teacher, 50:208-212, 1988.
21. Eisen, Y., Stavy, R.: Development of new science study unit following a research on students' ideas about photosynthesis: A case study in: Adolescent Development and School Science (P. Adey, ed.), pp. 295-302. East Susses, England: The Palmer Press.Lewes 1989.
22. Eisen, Y., Stavy, R., Barak-Regev, M.: Is is possible to study photosynthesis without misconceptions? Paper presented at ATEE meeting, Limmerick, 1990.
23. Friedler, Y., Tamir, P.: Basic Concepts in Scientific Research. Jerusalem: Israel Science Teaching Center, Hebrew University, 1990.
24. Friedler, Y., Amir, R., Tamir, P.: High School students' difficulties In understanding osmosis. International Journal of Science Education, 9:541-551, 1986.
25. Grobman, A. B.: The Changing Classroom. Garden City, NY: Doubleday 1969.
26. Gurley, L.: Gowin's Vee, The Science Teacher, 50-57, 1992.
27. Haslam, F., Treagust, D.F.: Diagnosing secondary students' misconceptions of photosynthesis and respiration in plants using a two-tier multiple choice instrument. Journal of Biological Education, 21:203-211, 1987.
28. Helm, H., Novak J.D. (Eds): Proceedings of the international seminar: Misconceptions in science and mathematics, Cornell University, Ithaca, NY, 1983.
29. Lively, W.: How much science does the public understand, American Scientist, 76(5), 439-443, 1988.
30. Hurd, P.D.: Biology education. In: What Research Says to the Science Teacher (P. Kelly, G. Schaefer, eds.), Vol. 3, pp. 12-32. Washington DC: National Science Teachers Association 1981.
31. Jimenez Aleixandre, M.P.: Los esquemas conceptuales sobre la Seleccion Natural: analisis y propuestas para un cambio conceptual. Doctoral Thesis. University Complutense, Madrid, 1990.
32. Jimenez Aleixandre M.P. Thinking about theories or thinking with theories? A classroom study with Natural Selection. Int. J. Sci. Educ., 14(1), 51-61, 1992.
33. Keeton, W. T. McFadden, C.H.: Elements of Biological Science. 3rd ed. New York: Norton 1983.

34. Kelly, P.J.: The structure of biological education. In: Biological Education for Community Development (P.J. Kelly, G. Schaefer, eds.), pp. 22-31. London: Taylor and Francis 1980.
35. Kelly, P.J., Wray, J.D. (Eds.): The Educational Use of Living Organisms – A Source Book. London: English Universities Press 1975.
36. Kimball, J.W.: Biology. 5th ed. Addison-Wesley 1983.
37. McInerney, J.D.: Human ecology: Restoring Life to the biology curriculum. In: High School Biology Today and Tomorrow (W.G. Rosen, ed.), pp. 117-124. Washington DC: National Academy Press 1989.
38. Novak, J. (Ed.): Second International Conference on Misconceptions and Teaching Strategies in Science and Mathematics, Ithaca, NY: Cornell University 1987.
39. Novak, J. E., Gowin D.B.: Learning how to learn. Cambridge, UK: Cambridge University Press 1984).
40. Novak, J.: Clarify with Concept Maps, The Science Teacher, 45-49, 1992.
41. Oram, R.F., Hummer, P.J., Smoot, R.C.: Biology: Living Systems.Merril, Columbus Ohio 1983.
42. Pfundt, H., Duit, R.: 1991 Bibliography: Students' alternative framework and science education. (3rd ed.) Kiel. IPN, 1987.
43. Posner, G.J., Strike, K.A., Hewson, P.W., Gerszog, W.A.: Accommodation of a scientific conception: Toward a theory of conceptual change. Science Education, 66:211-227, 1982.
44. Ray, G. C.: Coastal Zone Biodiversity Patterns, BioScience, 41:7, 490-497, 1991.
45. Roth, K.J., Anderson, C.W.: The Power Plant. Institute for Research on Teaching. Michigan State University, 1985.
46. Schwab, J.J.: Biology Teachers' Handbook. New York: John Wiley & Sons, 1963.
47. Short, F. K. et al: Eel grass wasting disease: cause and recurrence of a marine epidemic. Biological Bulletin, 173: 557-562, 1987.
48. Short, F. K.: An Estuarine Profile and Bibliography of Great Bay, New Hampshire , Jackson Estuarine Laboratory, Durham, NH, 1992.
49. Shymansky, J.A., Kyle, W.C.: A summary of research in science education – 1986. Science Education, 72: 249-253, 1988.
50. Silberstein, M., Tamir, P.: Questionnaire data as a source for curriculum planning. Studies in Educational Evaluation, 5:209-214, 1979.
51. Silberstein, M., Tamir, P.: Factors which affect students' attitudes toward the use of living animals in learning biology. Science Education, 65:119-130, 1981.
52. Simpson, M., Arnold, B.: The inappropriate use of subsumers in biology learning. European Journal of Science Education, 4 173-182, 1982a.
53. Simpson M., Arnold, B.: Availability of prerequisite concepts in biology learning. European Journal of Science Education, 16: 65-72, 1982b.
54. Smith, E.L., Anderson, C.W.: Plants as producers: A case study in elementary science teaching. Journal of Research in Science Teaching, 21: 685-698, 1984.
55. Stavy, R., Eisen, Y., Yaakobi, D.: How Israeli students aged 13-15 understand photosynthesis. International Journal of Science Education, 9:105-115, 1987.
56. Stewart, J.: Difficulties experienced by high school students when learning Mendelian genetics. The American Biology Teacher, 44 80-83, 89, 1982.
57. Stilwell, S., Brody M.: Cross cultural analysis of Canadian and Maine students understanding of the Gulf of Maine, Proceedings of the Second International Conference on Misconceptions of Science and Mathematics, Cornell University, Ithaca, NY, 1987.

58. Tamir, P.: Life science. In: International Encyclopedia of Education (T. Husen, T.N. Postlethwaite , eds.), Vol. 5, pp. 3056-3061. Oxford: Pergamon Press 1985.
59. Tamir, P., Hamo, A.: Attitudes of Israeli students to the use of animals in the biology classroom. International Journal for the Study of Animal Behavior, 1:299-311, 1980.
60. Tamir, P., Zolar, A.: Anthropomorphism and teleology in reasoning about biological phenomena. Science Education, 75:57-68, 1991.
61. Tamir, P.: Retrospective Causal Reasoning (RCR) in Biology. In Relations and Biology Learning: the Acquisition and Use of Knowledge Structures in Biology (K. Fisher, M. Kibby, eds.). Springer-Verlag.
62. Tamir, P., Jungwirth, E.: Botany and zoology: A curriculum problem. Journal of Research in Science Teaching, 11:5-16, 1974.
63. Tamir, P., Gal-Choppin, R., Nussinovitz, R.: How do intermediate and junior high students conceptualize living and nonliving? Journal of Research in Science Teaching, 18: 241-248, 1981.
64. Walker, D.: Foundations of Curriculum. San Diego: Harcourt Brace Jovanovitz 1990.
65. West, L.H.T., Pines A.L., (Eds.): Cognitive Structure and Conceptual Change. Orlando, FL; Academic Press 1985.
66. Westbury, I., Wilkof, N. (Eds.): Science, Curriculum and Liberal Education, The Chicago University Press 1978.
67. Wood-Robinson, C.: Young people's ideas about plants. Studies in Science Education, 19 19-135, 1991.

The Graphic Representation of Biological Knowledge: Integrating Words and Images

James H. Wandersee

Biology Education Program, Department of Curriculum and Instruction
223-F Peabody Hall, Louisiana State University, Baton Rouge, LA 70803

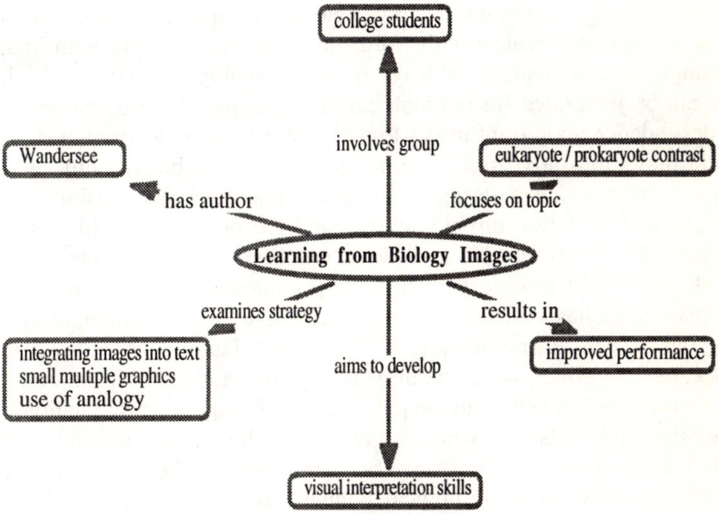

Abstract. The key question addressed in this study was: 'Can biology textbooks be improved so that students are more successful in integrating words and images?' This research indicates that learning from college biology textbooks can be enhanced by replacing traditional expository text accompanied by single, captioned, prototypical micrographic images with expository-narrative text integrated with small-multiple micrographic images. Performance on a posttest calling for identification of transmission electron micrographs of unknown cells as either prokaryotes or eukaryotes, circling the cell nuclei on unfamiliar micrographs depicting either prokaryotes or eukaryotes, and applying the town-city analogy to unfamiliar micrographs was used to assess learning outcomes. Students using the experimental approach outperformed those using a traditional textbook approach. The learning gains detected were predicted using Ausubelian learning theory.

Keywords. Biology education, concept learning, examples, scientific text comprehension, graphic representation, small multiples, electron micrographs, prokaryotes, eukaryotes

The Context

There is a growing awareness among bioeducational researchers that the graphic representation of biological knowledge is a domain of vast potential for the improvement of biology teaching. Much biological knowledge consists of spatial/temporal/functional relationships which lend themselves nicely to visual display [cf. Fisher, in this volume]. Because textbooks continue to constitute a vital linkage between the constructed knowledge base of biology and the students who are attempting to construct their own conceptual understanding of the discipline, and because biology textbooks are somewhat unique among science textbooks in that they are not only replete with words and diagrams but also with images, it seems imperative to search for ways in which biological text and biological images can be integrated for optimal learning. During a time of transition from heavy dependence upon print media to a mix of print and electronic media, such research may not only improve the quality of textbooks, but also help curriculum and courseware designers make sound educational decisions.

Upon what foundation might such research be built? In his little gem, *On Becoming a Biologist*, author John Janovy, Jr. observes that "the world of life is, above all, the world of objects...while we [biologists] endure abstractions, we prefer to *see* [emphasis added] things that are alive" [8]. In commenting on the graphic richness of an introductory biology text," Janovy concludes that "in a complex but tangible world, understanding often comes from combining observation of what is new with the physical act of graphic representation."

I and other biologists with whom I have shared this quotation resonate at once with it; for us, seeing has not only been the route to believing, but also to understanding how things biological connect. It is therefore troublesome to confront research which reports that, while 25% to 33% of an introductory college biology textbook consists of images and other graphic aids, students routinely ignore them [5, 12]. Perhaps we need to rethink our use of graphics in biology teaching so that it builds upon what we already know about how students learn science. I remain convinced that the graphic representation of scientific knowledge is a new frontier where substantial improvements in science education may be found [14].

Therefore, my research group decided to address this graphic bypass phenomenon with a view toward improving biology textbooks. More specifically, we asked: If much of biological knowledge has been constructed from microscope-based images, how can college biology students become more visually literate? We wondered: "Can the biology textbooks we use be improved so that they are more successful in integrating images and text?" A 1991–1992 grant from the National Science Foundation funded a research project entitled *Exploring Microstructures: Introducing Biology Students to the Images, Tools, and Applications of High-Tech Microscopy*. It offered us an opportunity to pursue these and related questions.

First, a little background information may be helpful. Some organisms (eukaryotes) have cells which contain a special structure called the nucleus and other organisms (prokaryotes) do not. There are other important differences between them, as well. However, the cells of all living organisms fall into one of those two categories. We decided to focus our research on how students learn the prokaryote-eukaryote distinction. In our own teaching experience, not only is the distinction poorly explained in most biology textbooks, often in brief textual increments separated by many intervening pages, but it is also crucial to understanding a number of key biological principles throughout biology--from classification, e.g. the five-kingdom system, to evolution, e.g., the endosymbiotic hypothesis. In addition, this learning task involves the use of transmission electron microscope TEM micrographs – – arguably the most difficult of all microscopic images for beginners to interpret [2].

We were aware that some scientists think that the five-kingdom system needs revision [1, 3, 7, 9, 10, 15] – on the basis of RNA sequencing studies which indicate that there were two prokaryotic lineages, eubacteria and archaebacteria, which appear to be no more closely related to each other than either is to the eukaryotes. We knew that some microbiologists are already using a three-domain system as a result, and were also aware that there is evidence which suggests that the eukaryotic line of descent is not of relatively recent origin, as biologists once thought, nor are prokaryotic cells ancestral to it. However, these current scientific issues were not raised in the latest introductory college biology textbooks which we examined, nor did we see the prokaryote-eukaryote distinction as becoming less pedagogically important should textbook content shift.

One of the aspects of the contemporary textbook's presentation of this bio-educationally important distinction which seemed most problematic to us was the use of a single pair of captioned, prototypical micrographs located outside the text stream. Based upon the work of graphical theorist Edward R. Tufte [16, 17], we decided to design a pair of small-multiple graphics (each consisting of five micrographic images carefully chosen to present a range of instances of both basic cell categories to the learner's eye) to replace the single pair of prototypical prokaryote and eukaryote cells commonly found in biology textbooks when this distinction is introduced.

Since learning theorist Joseph D. Novak and philosopher D.Bob Gowin [13], both influenced by the work of psychologist David P. Ausubel, contend that we think with concepts and that concepts are patterns in objects (or events) to which we assign a language label, we decided to use small-multiple graphics to help the learner identify and construct the visual pattern of difference between the two cell categories and then we attempted to help the learner attach meaning to those differences. Working from the aforementioned definition of a concept, it seemed theoretically sound to assume that providing a student with a single instance of a concept, e.g., a prototypical cell, is not enough to teach the pattern underlying the concept in a meaningful way. In addition, because comprehension of expository writing (typical of biology texts) requires greater attention to text structure, we introduced elements of narrative writing, e.g., an analogy based upon the differences between towns and cities, (suggested by Prof. M.D. Socolofsky – a member of the course team) in the text which we wrote to accompany our small multiples – with the aim of improving biology students' overall comprehension of the distinction.

Because eye-fixation studies [6] indicate that when students read texts accompanied by graphics, they first attempt to construct an understanding of the text and then test their understanding by using the graphic, we conjectured that if text and graphics were truly to become unitary, the placement and captioning of the graphics must become integral to the reading of the text – so that students could no longer use each component separately.

Thus, what differentiates our treatment of the prokaryote-eukaryote distinction from that of the typical biology textbook is: (a) the use of small-multiple micrographs to teach the perceptual differences central to the distinction, (b) the use of crisp expository text coupled with the introduction of narrative text devices, e.g., analogy, to first guide the discovery of and then attach meaning to those perceptual differences, and (c) the use of flanking, columnar graphics (in small-multiple format), along with text-embedded captioning, to draw the relevant electron micrographs into the text stream for first-pass processing by the reader.

Subsequently, we decided to compare the learning that occurred when college students used our own simulated textbook, 1-page excerpt explaining the prokaryote-eukaryote distinction with the learning that occurred using a functionally equivalent introductory 1-page excerpt drawn from a leading college biology textbook [4], which we considered to be pedagogically superior to the others we examined. See Appendix A to make a comparison of page design and content information.

The research question posed was: "Does the use of what Tufte [16, 17] calls small-multiple graphics, here two sets of five TEM micrographs providing a range of examples for each of the two categories, when combined with a crisp expository/narrative text and text-embedded captioning, and when executed with flanking columnar graphics, improve student learning of the prokaryote-eukaryote distinction, as compared to an equivalent traditional expository biology text accompanied by a pair of prototypical TEM micrographs?"

Method

Subjects

The college biology students (N = 237) who participated in this study, which compared the learning outcomes resulting from two textbook treatments experimental and traditional, of the prokaryote-eukaryote distinction, were all community college students enrolled in an introductory biology course. Six biology instructors chosen by the administration at the state's largest community college administered the research instruments following a stringent protocol. The college is located in a large metropolitan area in the South. The data collected represent a total of 39 hours of student time devoted to this research task. Students receiving the experimental textbook treatment, devised by us, were similar in number to those receiving the traditional textbook treatment (121 versus 116) and the ratio of males to females was approximately the same in each sub-sample. For the entire sample, approximately 25% of the students were male and 75% were female--reflecting both the program offerings of the college and the male-female enrollment ratio at the college.

Procedure

Students were assigned randomly by class to receive either the 1-page experimental textbook excerpt or the 1-page traditional textbook excerpt. Students were allowed 8 minutes to read and examine the excerpt they were given. It was then collected. Next, in a posttest-only design, students were allowed 12 minutes to complete a two-sided, single-sheet questionnaire designed to probe both the perceptual basis and their conceptual understanding of the prokaryote-eukaryote distinction. A total of 12 multiple-choice, short-answer, and cell identification/drawing tasks comprised the questionnaire. The instrument was pilot tested with several students, critiqued, and refined prior to large-scale administration. It is noteworthy that the questionnaire contained TEM micrographs which the students had not viewed previously. All of the TEM micrographs used in this study were taken by research team members and professional electron microscopist Cindy Henk, and were reproduced for this study from her research-quality images.

Scientists on the research team verified the categorizations of the cells pictured on the questionnaire. A three-category system for analyzing the open-ended questions was developed by several research team members. Categorization was achieved using standard qualitative practices and interrater agreement was calculated at 0.82, using the technique recommended by Krippendorf [11]. The coded results were analyzed via microcomputer, using relational database and statistical software.

Results

For the sample described previously (cf. Table 1), we asked the following question: How accurate were the students in identifying unknown cells as being either a prokaryote or eukaryote cell? Table 2 shows that the experimental treatment promoted greater learning for all students than did the traditional textbook approach. It should also be noted that, while both sexes benefited from the experimental approach, the experimental treatment resulted in a dramatic improvement in the performance of male students on cell-identification tasks related to the prokaryote-eukaryote distinction.

Table 1. Sample composition by treatment and gender

Text Treatment	*Female*	*Male*
Traditional (control)	85	31
Experimental	95	26
Totals ($N = 237$)	180	57

Table 2. Accuracy in cell identification task performance, by treatment and gender

Text Treatment	*Female*	*Male*
Traditional (control)	21%	6%
Experimental	27%	39%

Note. Each number represents the percentage of the group numbers who were able to perform all nine cell identification tasks successfully (100% accuracy). $N = 237$.

Table 3. Capability to apply the town-city analogy to actual micrographs, by treatment and gender

Text Treatment	*Female*	*Male*
Traditional (control)	39%	23%
Experimental	47%	81%

Note. The number for each group represents the percentage of students who were at least 80% successful in applying the analogy to five different micrographs. $N = 237$.

Table 4. Demonstrated accuracy in locating the nuclei present in a set of micrographs containing eukaryotic cells, by treatment and gender

Text Treatment	*Female*	*Male*
Traditional (control)	37%	42%
Experimental	46%	69%

Note. The numbers represent the percentage of students in each group who were 100% successful in locating the nuclei within a set of five micrographs containing some eukaryotic cells. Students circled the nuclei by marking directly upon the micrographs. $N = 237$.

We also sought to answer the question: How successful were the students in applying the town-city analogy to actual electron micrographs? Table 3 indicates that the experimental excerpt, in which the town-city analogy was elaborated, was more effective in promoting analogical interpretation than was the traditional text. Both males and females performed better using the experimental text, but again, the males' performance improved even more dramatically than did the females' performance. Since the ability to successfully use analogies is an index of meaningful learning--of substantive, multiple connections between concepts, these results are quite interesting.

To be certain that students were identifying the cells on the basis of relevant perceptual differences and not by extraneous features, we also asked students to identify the precise location of the cell nucleus in each member of a set of micrographs, if present, by free response (circling it on the micrograph). Table

4 shows the percentages of students who were 100% successful in locating the nuclei that were visible within a set of five TEM micrographs. By setting such high standards (100% success rate), we performed a stringent analysis of microstructure interpretation by treatment and gender.

Foremost, it is evident that the experimental treatment was superior for both groups. Once again, the performance of the male subjects improved more than that of females. However, here – for the first time – males outperformed females even when the males were taught using the traditional approach. In addition, on every table which we have examined, the males receiving the experimental treatment were the top performers.

Discussion

The use of single, prototypical cells to teach key biological concepts appears to be less effective than the use of small-multiple micrographs. The Ausubel–Novak–Gowin theory of learning [13] predicts this in its definition of a concept. Textbook publishers usually publish single, outstanding micrographic images which are atypical of those working biologists actually encounter. The author suggests that the lack of opportunity for students to compare images and then to construct the identifying criteria for themselves diminishes the likelihood that meaningful learning will occur. The small multiple graphic seemingly solves the comparison problem, is compatible with the emphasis upon signal differences inherent in visual processing of photographic images, and the five images making up the test material fit well with the known capacity of working memory.

Although the comparison was made between the traditional textbook approach and our experimental approach, it would be premature to assign the variance in performance to particular changes we made, e.g. graphic design, multiple images, altered text. From the free-response statements that students made on the questionnaire, our hypothesis is that the integration of carefully worded expository-narrative text and small-multiple images accounts for most of the improvement we noted.

While familiar with the literature on gender differences and visualization, we think it premature to speculate on the performance differential we observed in working with this population of students. Both males and females benefited from our approach and we intend to explore ways in which its constitutive variables may be changed to improve females' performance even more.

In studying how students learn from electron micrographs, we are exploring a content area where students bring little prior knowledge to the biology classroom. One does not normally encounter transmission electron micrographs in everyday life. Yet these images are crucial to understanding today's biology. By probing the structure of students' knowledge in this domain, we can examine science learning in a less confounding context. We think it important to relate these strange images to the high-tech tools that made them and to the applications of such images in science and society.

Published images taken with the TEM have extremely high data-ink ratio [16]. This makes them much harder for the beginner to interpret and understand than simpler graphics. One aspect of our research is now focused upon the selective degrading of TEM images (via image processing software), so that the step size of the microstructural topography is larger and visual differences are exaggerated for the benefit of the novice. One of the models that guided our work on integrating words and images was designed in 1613 by Galileo, who published the first telescopic observations of Planet Saturn. What was most striking to us was the way Galileo treated word and graphic as one. He put his sketches on the same lines as his words so that his images became elements in the text stream, to be read just like words [16]. Today's sophisticated word-processing software may make Galileo's idea – what was once a typesetter's nightmare – easy to accomplish. As biologists and bioeducators teaching specialized biological topics, this may be just the kind of text we need!

Finally, we are convinced that to learn the prokaryote-eukaryote cell distinction in a meaningful way, one must study actual electron micrographs. Using them, the learner must first be guided to make fundamental perceptual distinctions, e.g., presence or absence of a nucleus, relative size of cells, level of microstructural detail [compartmentalization], in descriptive fashion – without scientific labels. Once learners have come to recognize these visual patterns, they can build meaning by making connections to the relevant scientific concepts and principles. Then and only then can the names (concept labels) of structures and the functions of structures be learned in a meaningful way. Guided perception, concept formation, establishment of relations among and between concepts, development of higher-order concepts (constructs) – – all of these lead to key principles of cell biology. But there is no short cut! Simply telling a student those principles 'short-circuits' the learning process and invites rote learning. Like an amateur working on a home wiring project, making inappropriate or hasty connections can actually threaten rather than empower. Let novices be novices. There are no instant experts in biology.

Acknowledgments

This research was supported, in part, by NSF/LaSER grant (1991) HRD-01. The author, who was also the principal investigator, wishes to thank John St. Julien, Catherine Cummins, Cindy Henk, Marion Socolofsky, Sharon Matthews, Becky Demler, and John Lynn – all from Louisiana State University – for their valuable advice and assistance. The author also wishes to thank the biology faculty members at Delgado Community College in New Orleans, LA, who also played a vital role in gathering the data for this study . However, the views expressed in this chapter, as well as any errors or inaccuracies contained herein, are solely attributable to the author.

References

1. Brock, T. D., Madigan, M.T.: Biology of microorganisms (6th ed.). Englewood Cliffs, NJ: Prentice Hall 1991.
2. Burgess, J., Marten, M., Taylor, R.: Microcosmos. New York: Cambridge University Press 1987.
3. California Department of Education.: Science framework for California public schools – kindergarten through grade twelve. Sacramento, CA: California Dept. of Education Bureau of Publications 1990.
4. Campbell, N.A.: Biology. Menlo Park, CA: Benjamin/Cummings 1987.
5. Frazier, D.W.: Directed and spontaneous transfer of college developmental reading students' textmarking strategies. Unpublished doctoral dissertation, Baton Rouge, LA: Louisiana State University 1991.
6. Hegarty, M., Carpenter, P.A., Just, M.A.: Diagrams in the comprehension of scientific text. In: Handbook of Reading Research (R. Barr, M.L. Kamil, P.B. Mosenthal, P.D. Pearson, Eds.), Vol. II, pp. 652-653. White Plains, NT: Longman 1991.
7. Han, T., Runnegar, B.: Megascopic eukaryotic algae from the 2.1-billion-year-old Negaunee Iron-Formation, Michigan. Science, 257, 141, 1992.
8. Janovy, J.: On becoming a biologist. New York: Harper & Row 1985.
9. Kabnick, K.S., Peatie, D.A.: Giardia: A missing link between prokaryotes and eukaryotes. The American Scientist, 79, 34-43, 1991.
10. Kaufman, R.: Renowned microorganism researcher receives 117-year-old Dutch award. The Scientist, June 22, 23, 1992.
11. Krippendorff, K.: Content analysis: An introduction to its methodology. Beverly Hills, CA: Sage 1980.
12. Nist, S.L., Kirby, K.: The textmarking patterns of college students. Reading psychology, 10, 321-338, 1989.
13. Novak, J.D., Gowin, D.B.: Learning how to learn. New York: Cambridge University Press 1984.
14. Novak, J.D., Wandersee, J.H. (Eds.): Perspectives on concept mapping [Special issue]. Journal of Research in Science Teaching, 27(10), 1990.
15. Rivera, M., Lake, J.A.: Evidence that eukaryotes and eocyte prokaryotes are immediate relatives. Science, 257, 74-76, 1992.
16. Tufte, E.R.: The visual display of quantitative information. Cheshire, CT: Graphics Press 1983.
17. Tufte, E.R.: Envisioning information. Cheshire, CT: Graphics Press 1990.

Appendix A. Experimental Treatment

Eukaryotic Cells

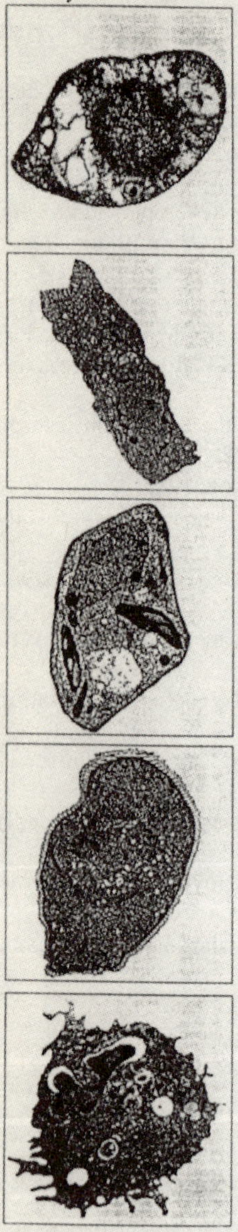

Prokaryotic Cells

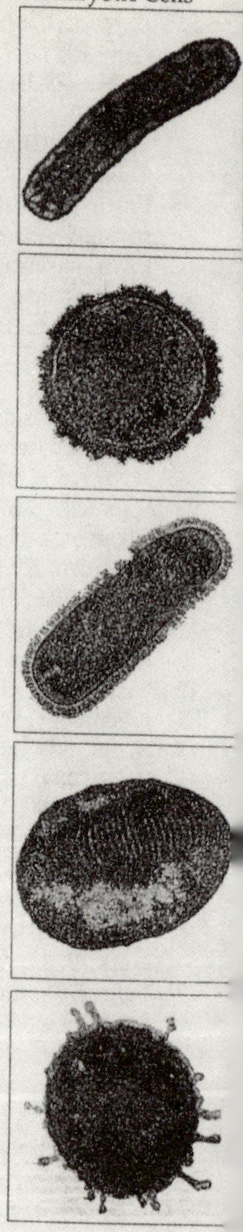

The Cell as a Community

The two groups of photographs on this page represent one of the most basic divisions of life -- cells with and cells without a nucleus. This division is so sharp that many biologists can recognize a cell as one type or another at a glance. Can you notice the differences?

All living organisms, including the simplest cell, can be regarded as communities. Like human communities, cells may vary in size and complexity. Each cell on the left can be thought of as a large city. Cities are divided into areas that perform specific functions like fire departments, power plants, water systems, and communication systems. In general, these cells are large and have distinct parts called organelles (little organs). Each type of organelle has a specific function for the cell as each area in the city has a specific function. You can see these organelles as distinct bodies in the pictures on the left. The largest of these organelles is the nucleus, which appears as a dark area with smaller dots within it. Can you find the nucleus in each picture on the left? The presence of this nucleus gives this type of cell its name, eukaryotic (*eu* = true and *karyon* = kernel [nucleus] in Greek).

Each cell on the right can be thought of as a small town. In small towns, many people have more than one job in the community and there is less division of labor into specific areas. In general, these cells are small, simple, and show fewer distinct bodies within the cell. Compare the cells on the left with those on the right to see if you can detect this difference. The absence of the nucleus gives this type of cell its name, prokaryotic (*pro* = before and *karyon* = kernel [nucleus] in Greek).

From the differences visible in these electron micrographs flow much of our understanding of the evolutionary history of life and some of the evidence for the 5 Kingdoms into which we divide all life. Prokaryotic cells are found only in Kingdom Monera while eukaryotic cells are found in the four remaining kingdoms. To review, how does the story about cities and towns relate to one of the most basic divisions of life?

Appendix B. Traditional Treatment

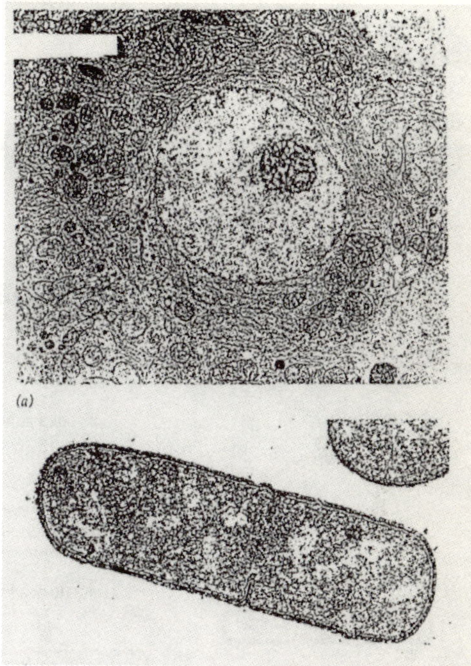

(a)

(b)

Figure 1.4
Two types of cells, as viewed with the electron microscope. *(a)* The eukaryotic cell, found in plants, animals, and all other organisms except bacteria, is characterized by an extensive subdivision into many different compartments, or organelles. The large spherical organelle in the center of this cell is the nucleus, which contains a smaller sphere called the nucleolus. Various organelles can be seen in the cytoplasm outside the nucleus. *(b)* The prokaryotic cell, unique to bacteria, is much simpler, lacking most of the organelles found in eukaryotic cells.

The Cellular Basis of Life

Two major kinds of cells can be distinguished based on structural organization (Figure 1.4). The eukaryotic cell, by far the more complex, is subdivided by internal membranes into many different functional compartments, or organelles, including a nucleus and the various cytoplasmic organelles. The prokaryotic cell is much simpler in organization. Its DNA is not separated from the rest of the cell into a nucleus, and most of the other organelles typical of eukaryotic cells are also lacking. The cells of the microorganisms known as bacteria are prokaryotic. All other forms of life are composed of eukaryotic cells.

Although eukaryotic and prokaryotic cells contrast sharply in structural organization, they have many similarities, especially in their chemical processes. Among eukaryotic cells, there is extensive variation in size, shape, and specific structural features, but certain organelles and process are remarkably similar form one cell type to another. For the most part, cellular diversity represents variations on a common theme.

Components of Comprehension Monitoring in the Acquisition of Knowledge from Science Texts

José Otero

Departmento de Física
Universidad de Alcalá, 28871 Alcalá de Henares, Madrid, Spain

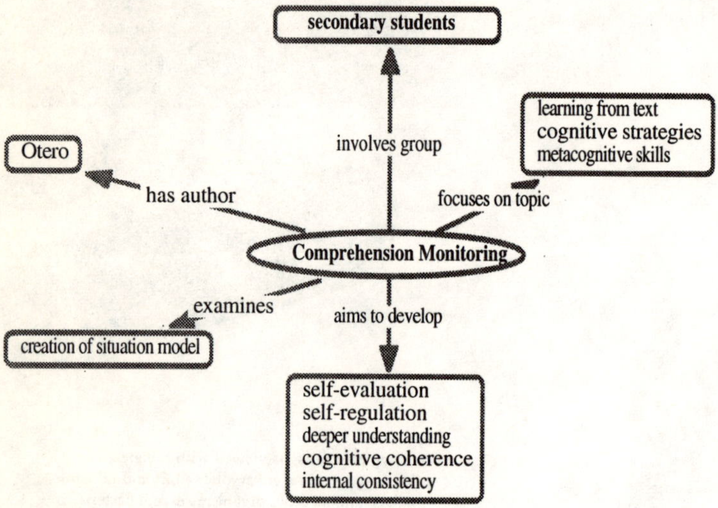

Abstract. The chapter analyzes the strategy of comprehension monitoring when acquiring knowledge from science texts. Also, secondary school students' failures in comprehension monitoring are identified in terms of its components processes. Two phases are distinguished in the monitoring of comprehension: the evaluation phase--knowing if one understands--and the regulation phase – taking action to repair identified problems. Science students may fail in evaluating comprehension because of the use of inappropriate cognitive strategies or inadequate comprehension standards. The problems identified in the regulation phase are related to low levels of acceptable coherence in the knowledge structures created by the students.

Keywords: comprehension monitoring, metacognitive strategies, metacognition, science texts, science learning

The well known motto of Ausubel, "The most important single factor influencing learning is what the learner already knows," has influenced the activity of science education researchers in the past fifteen years. Ascertaining the knowledge which students already have in the form of preconceptions or misconceptions and trying to 'teach accordingly' has taken time and effort of science teachers and researchers. However, there are other factors that influence the acquisition of scientific knowledge. More than twenty years ago, E.Z. Rothkopf, who coined the term 'mathemagenic activities' to mean behaviors which give birth to learning, suggested:

> Consider how students learn from written material. What determines which capabilities a student has acquired after exposure to an instructional document? The content of the instructional material is undoubtedly important. So is, to a lesser degree, its organization. But most important, to an overwhelming degree, is what the student *does with* the instructional document. In a sense the student has complete veto power over learning, since without some activity on his part the instructional objectives can never be achieved. [15, p. 326] (emphasis in original)

Thus, one other factor influencing the acquisition of scientific knowledge is what the learners do to learn, the strategies they employ. Many such strategies have been studied, both at the cognitive and metacognitive level (see, for example, [3, 20]). Comprehension monitoring is one of these metacognitive strategies; it is used to decide when one understands or does not understand, and to act accordingly. The strategy of comprehension monitoring when acquiring knowledge from science texts is analyzed in the following, identifying secondary school students' failures in using its components processes.

Short introductory physics texts were used in the research reported here. However, the comprehension monitoring processes which have been identified will probably apply equally in other scientific disciplines, in particular when students acquire biological knowledge from textbooks.

Evaluation of Comprehension

Two phases are distinguished in the monitoring of comprehension although they are not entirely independent. First there is the evaluation phase: knowing if one understands or does not understand. Students may be unaware of their lack of understanding as a result of inadequate cognitive strategies. Take for example someone who reads a text but does not use the structure strategy [9], that it does not attempt to look for relations among the most general ideas in the text. The evaluation of comprehension would be especially inadequate in situations where this top-level structure plays a crucial role, as in a mathematical proof. This is illustrated by the following example. In a pilot study [13], several 18-year-old physics students, in the last year of Spanish secondary education, in groups of three, were given half a page of text consisting of an incorrect Proof of Newton's

Second Law involving a circular argumentation. Newton's Second Law, $\mathbf{F} = m\mathbf{a}$, was stated at the beginning as a foundation for arguments used to 'deduce' $\mathbf{F} = m\mathbf{a}$ at the end. The low level content of the text was chosen so that, according to their teacher's criterion, it would be understandable for these students who had been studying physics for three consecutive years. The students read the text and, were asked to point out any difficulty which they might have found. Some groups of students did not notice any problem in the text. For example, a student in one of these groups, when asked to recall the text, wrote:

> Newton's Second Law: F = ma
> A force is applied, F = 0
> Thus: ma = 0
> This finally proves that the force exerted on a body
> equals its mass times the acceleration.
> F = ma

Not identifying the top level relation in the text makes these students unaware of their lack of understanding. Evaluation, the first cognitive process involved in identifying a problem in the information to be processed, involves using various comprehension criteria [2]. Subjects who only pay attention to isolated meanings of the words or to local coherence of the text use limited criteria, thus hindering a proper evaluation of the text.

In another study of comprehension monitoring when reading science materials [14], six short texts were given to 12th grade students (18 years), and also to 10th grade students (16 years). Four of the texts (#2 'Acid rain', #3 'Superconductivity', #5 'Neutrinos', #6 'Ceramics') involved an explicit contradiction like that found in the following:

Text #5 Neutrinos

Neutrinos are particles with nearly zero mass. Their detection is very difficult because they do not react to magnetic or nuclear forces. In order to detect them a great amount of water is necessary, placed in a deep place underground, where it could be free from other radiations. A great amount of water is necessary because neutrinos seldom interact with matter. Several countries have set up neutrino detectors which will be useful in the future. The great facility with which neutrinos are detected makes them very suitable, for example, for the study of several cosmic phenomena.

The students were instructed to read the texts and report any difficulty which they may have had in understanding them. This had to be done by underlining the problematic words or sentences and explaining the nature of the difficulty in the space provided. In addition, they had to rate the understandability of the text on a four-point scale. In a second phase of the experiment they were informed of the existence of the contradictions in the text and were asked what they had done about them (details of the procedure can be found in [14]). It was found that 49%

of 10th grade students and 27% of 12th grade students did not notice the contradiction in any of the four texts and rated the understandability as good. These students evaluated their comprehension inadequately. The causes for this phenomenon may be faulty cognitive processes. In the Neutrinos text, for example, 47% of the students in both grades who did not detect the problem, recalled the first contradictory proposition (obviously without recalling the second), while only 30% recalled the second. A similar pattern of results was obtained for the other analyzed text, Superconductivity. This suggests that a possible mechanism for not being aware of the contradiction is the suppression of the second contradictory proposition, in the final representation of the text in the reader's memory. The Construction-Integration model of text comprehension [7] predicts exactly this [12]. Some of the readers place so much emphasis on the first contradictory proposition that it actually suppresses the second which it contradicts. Of course, not being aware of one of the contradictory terms leads to an inadequate evaluation of comprehension.

Information found in science textbooks does not involve contradictions like that shown above. However, the evaluation problems identified in this artificial setting could very well reflect those which students may have in more realistic learning situations: new information incompatible with what the learner knows can be overridden by old knowledge (see [1] for similar results).

Regulation of Comprehension

The previously described processes concerning the evaluation phase are cognitive. Metacognitive processes take place in the regulation phase. A prerequisite for regulation is being aware of a comprehension problem, that is, adequately evaluating comprehension. Once a comprehension problem is detected two courses of action are taken by the subjects who read the contradictory texts, according to their own reports [14]. Some of the subjects declare having noticed the problem but not having done anything to solve it. The comprehensibility rating given by these subjects to the text is high. The explanations given by these students in a subsequent interview suggests an inadequate use of comprehension criteria as an explanation for their behavior:

> **Interviewer:** When would you say that a text is difficult to understand?
>
> **Student 2024:** For instance ... I dunno ... when you use very complicated words or sentences which are a bit difficult ... well you understand them but they aren't normal ... sometimes some books use ... well ... weird words, ones you have never seen in your whole life.

In this case, it seems that a violation of the internal consistency standard is regarded as unimportant, compared to lexical or syntactical criteria [2]. There are other students who reported having "solved" the comprehension problem by an otherwise sensible procedure, generating inferences. Thus, in the Neutrinos text, a

10th grade student explains why having found the contradiction when reading the text, he had not explicitly pointed out the problem:

> **Interviewer:** In the one about neutrinos ... you realized that the sentences were contradictory, and then you did not underline them. Why not ?
>
> **Student 2006:** It says basically the same there ... I mean ... well it used to be difficult and then in some countries they have discovered these apparatuses that can detect those trivial values ...
>
> **I:** But you said that you realized that the two sentences were contradictory, what was the problem that you found?
>
> **S:** That they were difficult [to detect] before and then easy, and then I read it twice and I said that ... I mean that with those apparatuses which they mention, it should be easy.

The comprehension process for any text involves creating a representation, the situation model which includes text propositions and those retrieved from the reader's own knowledge base [18]. The latter includes bridging inferences, such as those necessary to understand anaphoras and elaborative inferences, like those that relate text information to the reader's own knowledge. Building a situation model for a text involves creating a coherent representation using the aforementioned elements. Both in terms of symbolic schema theories [4] and connectionist theories [16] this constructive activity has been described as a constraint satisfaction process aimed at determining

> ... which model best fits the incoming information. Ultimately, consistent configurations of schemata are discovered which, in concert, offer the best account for the input. This configuration of schemata together constitutes the interpretation of the input. [16, p. 18]

A number of propositions can be retrieved from the knowledge base of the reader to obtain this model, but not all of them make an equal contribution to the coherence of the result. Thus, regulation can be regarded as an activity aimed at obtaining the maximum coherence of this resulting configuration. This is depicted in Figure 1 using a propositional representation of meaning based on [6] and [17], where the information involved has been simplified. The text propositions involved in the contradiction are identified by P's while the repairing inference generated by the reader is identified by II. Links between propositions are established by the default criterion of argument overlap [18]. Student 2006, whose explanation is transcribed above, finds more coherence in a network like the one shown in Figure 1b rather than the one in 1a, although both representations contain incoherent propositions (connected by negative links). In addition, for this student the coherence of network 1b falls within her 'acceptable coherence level', since she did not reject the configuration as unacceptable. This acceptability level seems to depend on the reader's personal characteristics, and most probably on additional factors like task demands [8]. One can speculate that subjects who rejected the Neutrinos text as difficult to understand because of the contradiction, (17% in 10th grade and 64% in 12th grade) probably tried to make the same repairing inference and discarded the resulting configuration as inappropriate.

Examples can be readily found of the low coherence levels that are acceptable for some students who acquire knowledge from science texts. In an experimental situation similar to the ones mentioned above, students read a short text which contained the following contradiction: "The speed of light is the greatest which can be achieved. No object can travel at a speed exceeding that of light because its mass would become infinite ... There are detailed observations of the behavior of some objects when they surpass the speed of light". One 18-year-old, 12th grade student, having noticed the contradiction without pointing it out explains why the text is acceptable for her: "I thought that it was something scientifically established and that there were no errors" [11, p. 373]. The example suggests that increasing the value of the acceptable coherence level of the knowledge structures created by the learner, an important parameter in the regulation process, may be a fundamental prerequisite in the acquisition of knowledge by some students.

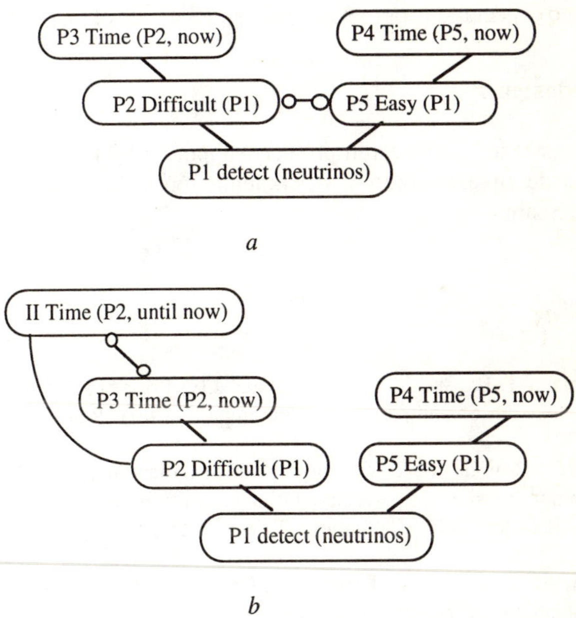

Figure 1a,b. Alternative propositional structures considered by reader 2006 (P's are text propositions, while II corresponds to an inference. Positive connections between propositions are indicated by arrows and negative connections, between incoherent propositions, by circles.)

Conclusions

Comprehension monitoring when acquiring scientific information from a text determines the final result of learning. The role of 'what the learner already knows' depends on comprehension evaluation and regulation processes. These are most important in cases where new information is seen as incompatible with old knowledge, as sometimes happens in science learning.

Both evaluatory and regulatory components should be taken into account when trying to improve comprehension monitoring. Students acquiring knowledge from science texts sometimes use limited comprehension monitoring standards in the evaluation phase. For example, they may check the text for lexical adequacy only and not for internal consistency. Improving the evaluation component requires improving reading strategies leading towards appropriate use of comprehension monitoring standards and deeper processing of the material [5, Chapter 6]. Improving the regulation component points to the importance of paying attention in science teaching to the level of coherence and comprehensibility accepted by students. This, in its turn, seems to be related to variables pertaining to areas that overlap both the metacognitive and the motivational fields [19,10], such as performance expectations or metacognitive judgments of one's own performance.

Acknowledgments

This paper is based on research projects supported by DGICYT (PS89-034) and the Centro de Investigación y Documentación Educativa of the Ministry of Education, Spain.

References

1. Alvermann, D.E., Smith, L.C., Readence, J.E.: Prior knowledge activation and the comprehension of compatible and incompatible text. Read. Res. Quart., 20, 420-436, 1985.
2. Baker, L.: How do we know when we don't understand? Standards for evaluating text comprehension. In: Metacognition, cognition and human performance (D.L. Forrest-Pressley, G.E. MacKinnon, T.G. Waller, eds.). New York: Academic Press 1985.
3. Brown, A., Bransford, J., Ferrara, R., Campione, J.: Learning, remembering, and understanding. In: Handbook of child psychology (J.H. Flavell, E. Markman, eds.). New York: Wiley 1983.
4. Collins, A., Brown. J.S., Larkin, K.M.: Inference in text understanding. In: Theoretical issues in reading comprehension (R.J. Spiro, B.C. Bruce, W.F. Brewer, eds.). Hillsdale, N.J.: Lawrence Erlbaum Assoc. 1980.
5. Garner, R.: Metacognition and reading comprehension. Norwood, NJ: Ablex Pub. Co. 1987.
6. Kintsch, W.: The representation of meaning in memory. Hillsdale, NJ: Lawrence Erlbaum Assoc. 1974.

7. Kintsch, W.: The Role of knowledge in discourse comprehension: A Construction-Integration model. Psych. Review, 95, 163-182, 1988.
8. Markman, E., Gorin, L.: Children's ability to adjust their standards for evaluating comprehension. J. of Educ. Psych., 73, 320-325, 1981.
9. Meyer, B., Rice, G.E.: The structure of text. In: Handbook of Reading Research (P.D. Pearson, ed.). New York: Longman 1984.
10. Nolen, S. B.: Reasons for studying: Motivational Orientations and Study Strategies. Cognit. and Instruc., 269-287, 1988.
11. Otero. J.: Comprehension monitoring in learning from scientific text. Proceedings of the Second International Seminar on Misconceptions and Educational Strategies in Science and Mathematics (J.D. Novak, ed.), pp. 370-375. Ithaca, NY: Cornell University 1989.
12. Otero, J., Kintsch, W.: Failures to detect contradictions in a text. What readers believe vs. what they read. Psych. Science, 3, 229-235, 1992.
13. Otero, J.: El aprendizaje receptivo de las ciencias: preconcepciones, estrategias cognitivas y estrategias metacognitivas. Tarbiya, 1(2), 57-66, 1992.
14. Otero, J., Campanario, J.M.: Comprehension evaluation and regulation in learning from science texts. J. of Res. in Sci. Teach., 27, 447-460, 1990.
15. Rothkopf, E.: The concept of mathemagenic activities. Rev. of Educ. Res., 40, 325-326, 1970.
16. Rumelhart, D.E., Smolensky, P., McClelland, J.L., Hinton, G.E.: Schemata and sequential thought processes in PDP models. In: Parallell distributed processing. Explorations in the microstructure of cognition. Vol 2: psychological and biological models (J.L. McClelland, D.E. Rumelhart & the PDP Research Group). Cambridge, MA: MIT Press 1986.
17. Turner, T.A.: The Propositional Analysis System. Version 1.0. University of Colorado, Institute of Cognitive Science, ICS Tech. Rep. 87-2, 1987.
18. Van Dijk, T., Kintsch, W.: Strategies of discourse comprehension. New York: Academic Press 1983.
19. Weinert, F.E.: Introduction and overview: Metacognition and motivation as determinants of effective learning and understanding. In Metacognition, motivation, and understanding (F.E. Weinert, R. H. Kluwe, eds.). Hillsdale,NJ: Lawrence Erlbaum Assoc. 1989.
20. Weinstein, C.E., Mayer, R.: The teaching of learning strategies. In: Handbook of research on teaching. 3rd. ed. (M.C. Wittrock, ed.). New York: Macmillan 1986.

Constructive Learning from Texts in Biology

Marlene Scardamalia[1], Carl Bereiter[1], Jim Hewitt[1], and Jim Webb[2]

[1] Ontario Institute for Studies in Education
[2] Huron Street Public School

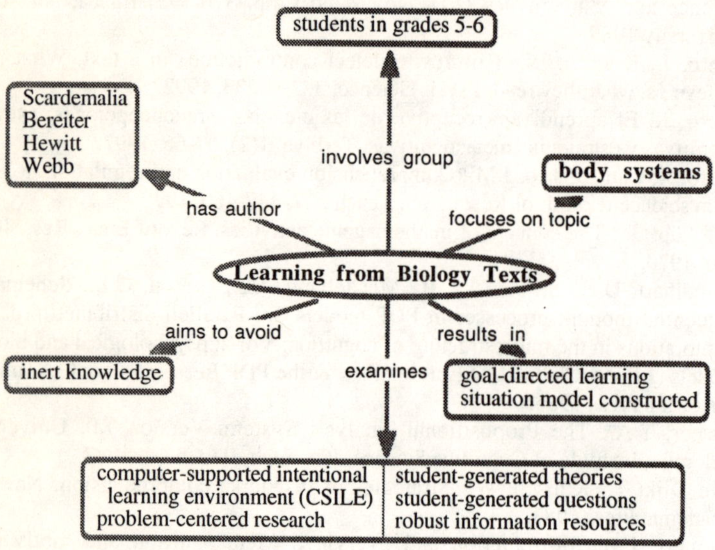

Abstract. In schools, discourse usually plays a part in the constructive process. We have been developing and experimenting with Computer Supported Intentional Learning Environments (CSILE) that combine the educational advantages of collective discourse with the tactical advantages of individual written and graphic work. In this paper we track three successive classes of grade 5-6 students working on the same biological topic (body systems). We find that students take a more goal-directed and constructive approach to using text information: (1) when students create a context for figuring out how things work by advancing their own provisional theories in advance of taking information from texts; and (2) when students identify gaps in their own knowledge.

Keywords. Constructive learning from text, biology learning, inert knowledge, problem-centered knowledge, situation model, CSILE, collaborative learning, problem centered student research

Passive learning from texts is often lamented. The typical elementary school biology text presents information in a topic-detail organization that is efficient for assembling and memorizing facts, but that is relatively useless for the pursuit of understanding. Students who passively take in what the texts present may end up with inert knowledge, difficult to apply and rapidly forgotten. Their actual understanding of biology, which is to say the problem-centered knowledge they call upon to make sense of the world and of stories and discourse in the real world, tends to be based on informal experience and hence deviates in many ways from scientific knowledge. As a reaction against such passive learning, many educators favor 'hands-on' methods which have weaknesses of their own [6], but which in any event cannot cover large parts of the terrain of biological knowledge. Therefore, the production of more active and constructive learning from authoritative sources remains a significant goal for instructional research.

What constitutes constructive learning from texts? It means more than constructing a textbase – a macropropositional representation of what the text says [4, 9]. It means bringing one's own mental models to the text, using them to identify significant text information, and revising one's mental models through efforts to reconcile model and text. This we take to be part of what Kintsch means by construction of a situation model, as distinct from constructing a textbase. In such cases it is the organization of the situation model – the model of that aspect of the world addressed by the text – rather than the organization of the text itself which dominates. Assuming that the students' mental models have been organized in ways that facilitate practical action and making sense of the world, they are likely to provide a better framework for organizing new information than the topic-detail framework typically provided by texts. We suggest that this is true, even if the student's mental model is naive, simplistic, even seriously in error. It at least provides a way of bringing new information obtained from texts into the process of understanding, and in this way there is the possibility of improving, perhaps even restructuring the mental model.

In schools, learning from texts is seldom a solitary endeavor. Discourse usually plays a part, even if it is only the recitation characteristic of conventional pedagogy. Whether discourse is of this ritualistic kind or whether it involves a more natural give-and-take of ideas, school discourse is likely to focus on what the text says, on achieving a consensual representation of its content [3]. Thus, in Kintsch's terms, it is concerned with constructing a textbase. According to the considerations set out in the preceding paragraph, however, it would be educationally preferable to focus instead on the students' own mental models and on the implications that text information has for them.

Although many educators might endorse this proposal as being in accord with their constructivist and child-centered philosophies, they would likely find it difficult to implement. How are the individual mental models of 30 or so students to be brought out for discussion and used as frameworks for considering the significance of text information? Daunted by the tactical difficulties of doing any such thing through the medium of class discussion, teachers are likely to resort instead to having the students do individual 'projects', in which they produce documents based on their own reading and deliberation. But such projects are famous for inducing passive use of text material – copying or paraphrasing text material and placing it into an organization much like the organization of the book or encyclopedia article it came from.

Computer Supported Intentional Learning Environments

The computer-supported learning environment, that we have been developing and experimenting with over the past 6 years, may be seen as an effort to combine the educational advantages of collective discourse with the tactical advantages of individual written and graphical work. Computer Supported Intentional Learning Environments (CSILE) is a networked system built around a communal database into which students enter text and graphic notes (Figure 1). Stored information is made public by default so that children may easily examine each other's contributions and work with one another. Access to database materials is provided by a flexible retrieval utility that permits complex searches by author, topic, and other attributes. Through this mechanism, students share their learning goals, theories, wonderings and discoveries.

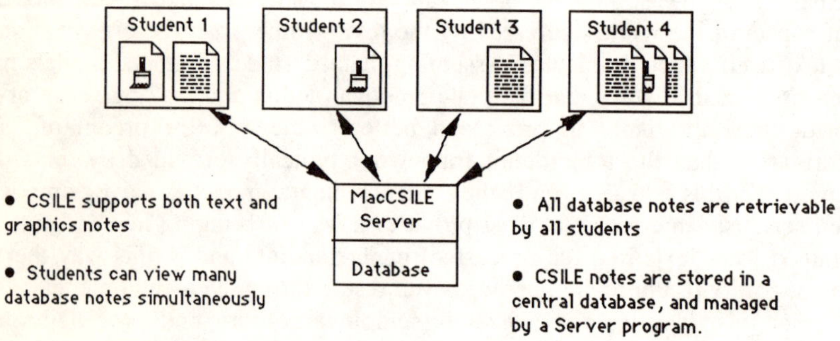

Figure 1. CSILE: A networked system for sharing text and graphics notes

A typical collaborative episode is shown in Table 1. Here, one grade 5 student makes a general inquiry about cells. A second student, seeing the question on the database, responds with helpful information.

Table 1. A typical collaborative episode in CSILE

Problem: Do cells keep you alive? – Julie.
Dear Julie: I think I can help you. Nothing can live without cells, and if something doesn't have cells, it is not a life form. – Cheryl.

The CSILE program allows students to move easily between individual and collaborative work. At an individual level, the database serves as a written record that allows students to return to ideas, set goals, construct plans, list questions, and record hypotheses. These writings are then made available to everyone, and become the framework of collaborative discussion. In this way, each learner is provided with the opportunity to describe their personal mental models, which in turn feed the knowledge-building activities of the classroom. This has several advantages over traditional groupwork activities. In comparisons of CSILE to face-to-face groups, we found that student contributions are quantitatively more equal over CSILE than when class discussion is their sole means of communication [2]. All participants have equal opportunity to express their ideas, and all participants have an equal voice. We believe this not only benefits the individual student, but it exposes the class to a greater variety of ideas than they would normally encounter in face-to-face situations.

The technology itself is neutral as to what students do with text information and with their own mental models. At one extreme, CSILE may be used simply as a medium for doing 'projects' of the conventional kind, with the benefit of text editing, color graphics, and storage and retrieval functions. At the other extreme, through its commenting functions and capabilities for collaborative building of networks of charts, it may be used for sustained collaborative inquiry. In this paper we shall report and examine a progression that took place over three successive years in the way grade 5–6 students worked on a biology topic in CSILE. The classes did not start at one extreme and move to the other, but the Year 1 class showed much of the behavior that has been associated with conventional 'projects'. The Year 3 class shows a decided shift in the direction of more collaborative inquiry, focused on their own mental models, their own questions and efforts to make sense of biological phenomena, rather than on what the available texts happened to present. The changes in procedure that brought this shift about were relatively minor, but apparently powerful in effect, and part of our effort in this paper will be to understand how they worked.

Method

We track, for three successive grade 5–6 classes taught by the same teacher, work on the same topic. Each year there was a common structure to what went on in the classroom. Approximately 30 grade 5–6 students each year received a brief introduction to the major systems in human biology: circulatory, immune, skeletal, respiratory, reproductive, and nervous systems. They were told to form themselves into groups and identify topics for research. The overview might be considered an introduction to the vocabulary of human biology, clustered more or less in the manner of a table of contents of popular curriculum material (see Table 2). Student work was entered into the CSILE database in the form of text or graphic notes, where it could be commented on by other members of the working group as well as by other students in the class.

Table 2. Topical organization of curriculum material

Source	*Topics*
Harcourt, Brace, Javanovich, 1980	Animal Structure and Function: Digestion, Respiration, Transportation, Excretion, Coordination, Reproduction; Cell Structure and Function and Heredity
Suzuki, David (1987); Toronto: Stoddart Publishing Co.	Looking at Looking at You, Your Skin, the Body. Your Bones, Your Muscles, Your Heart and Blood, Your Lungs, Your Digestive System, Your Brain and Nerves; How You Communicate

Year 1 Observations

In Year 1, nine student groups were formed, with two to five students per group. The nine groups selected topics as indicated in Table 3. As the table indicates, student inquiry focused on loosely structured ideas held together by broad topics of the sort found in curriculum materials. Students worked out among themselves the various topics that different students in the group would pursue. Figure 2 shows the self-titled, self-designed 'Respiratory System' group's plan of study, a rather disjointed array of issues. This approach was typical.

Table 3. Examples of student-selected research topics, year 1

The Skeletal System: Group of 3 students The spinal cord, the rib cage, the foot, the thigh, the knee, bone, the hand
Cells: Group of 3 students The nucleus, chromosomes, genes, nerve cells, muscle cells, white blood cells, skin cancer
The Respiratory System: Group of 4 students The lungs, smoking, origin of cigarettes, the mouth, asthmatics, the heart, respiration, asthma
The Senses: Group of 3 students The ear, the nose, the eye, blindness, deafness, nasal congestion, nosebleeds, odor

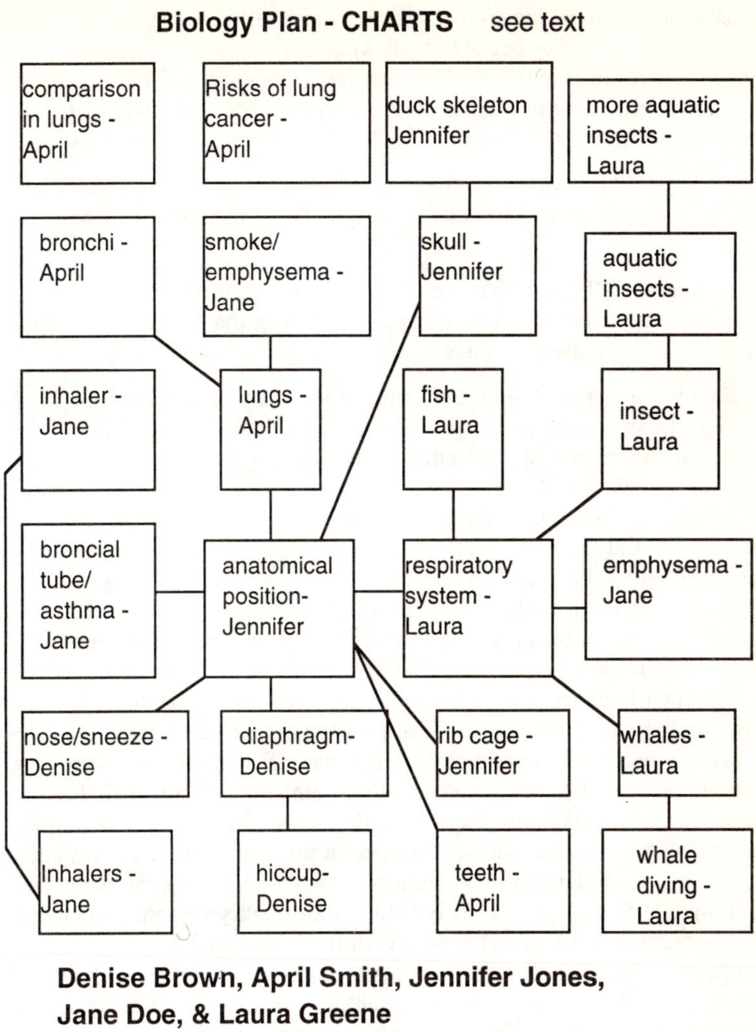

Figure 2. Respiratory group plan of study, year 1

Each group took a topical or structural rather than problem-centered approach to their study. A review of biology textbooks suggested that their models for how to divide up the world of ideas in human biology were based fairly directly on the books they used for reference. The books we reviewed show a common pattern, that we refer to as the topical/structural approach to learning human biology: chapters divide into broad topics of the sort suggested to students in the brief introduction to the unit, and within these chapters there are more- or less-detailed accounts of the organs associated with the various body systems, along with explanations of the workings of these organs and some suggestion of interrelationships between body systems.

When asked to set out issues that they wondered about, clarifying what they knew already and what they needed to find out to advance their understanding, we found students claiming to know facts of the sort set out in their textbooks:

> **Muscles:** "I know that muscles are made up of muscle tissues, and tissues are made up of muscle cells. Muscular dystrophy is the number one disease of muscles."

> **Endocrine System:** "I know that the endocrine system produces hormones...the parathyroids and thyroids are organs in the endocrine system and both help make hormones."

And when they wondered about things, they tended to wonder about the kinds of things you go to an encyclopedia or textbook to find out. For example, "What are the organs of the respiratory system?" "How are bones formed?" "What is the axial skeleton?" Generally, in Year 1, inquiry driven by student interests paralleled textbook approaches to the study of biology. It makes sense that their models for how to study are based on textbook models, since they are the only models they are likely to have had contact with.

Over the course of their study there was minimal integration of concepts. A common strategy was to collect a set of facts from an authoritative source and then reproduce that information in CSILE. On only a few occasions did students try to use their own knowledge as a basis for their research or attempt to make sense of how biological systems work. Their most productive research episodes seemed to occur when they stepped outside the static formulation of issues and topic groupings presented in textbooks. At this point they (a) identified novel, interesting research contexts for their investigations, (b) engaged in inquiry based on everyday phenomena, and (c) focused on problem-centered accounts of phenomena, trying to understand how things work. Such occasions were rare, but the resulting activity suggested the benefits of an inquiry-driven, knowledge-centered approach, which we experimented with in Years 2 and 3.

Some students in year 1 generated inquiries that were unlike the standard text-based questions that dominated the class database. These "I wonder" questions had the potential to engage students in novel, interesting research studies that would significantly advance their understanding of biological processes. About a dozen such questions were identified during the Human Biology unit. Examples:

> **Muscles:** "What's happening when you have a stiff muscle?" "What would happen if you didn't have muscles?"

> **Endocrine System:** "What does the endocrine system use to make hormones?" "What types of defects afflict hormones?"

> **Diseases:** "How do the medications for skin cancer work?"

Finding answers to the questions in the research materials was difficult, precisely because these questions were wonderment rather than text-based [7]. Consequently, students often identified interesting research problems, but had trouble advancing on them. Some students recognized that the textbooks were not providing them with all the information they needed. For instance, one child working on the endocrine system followed up his question-asking session by entering 7 notes into CSILE's database. Notes were titled: Pituitary gland, pancreas, adrenal glands, thyroids, parathyroid, kidney, and thymus. Material seemed to be copied from textbooks. Then he wrote the following note:

> My answer to the question how endocrine organs know when to produce hormones is that the pituitary gland sends hormones which stimulate the other organs such as the adrenal glands to create hormones, *however, I don't know how the pituitary gland knows when to produce hormones.* My answer to what hormones are made of is, hormones are made of organic material*(my source did not say what type of organic material).* My answer to why we need hormones is we need hormones to regulate our body and to maintain homeostasis. My answer to the question can defects in the endocrine system be environmentally caused or caused by genes that were passed on or both is both.

Here we see the student's original questions played an important role in shaping his research. We have italicized statements showing that the student understands that textbook material does not provide all the necessary information, and that there remain gaps in his understanding. Questions like these motivated the development of an "INTU" (I Need To Understand) facility in Year 2 for identifying knowledge gaps, and bringing these gaps forward for progressive refinement. A parallel and integrated NI (New Information) facility was added that encouraged students to record new information that they found in the course of their investigation. We stressed 'new information' rather than 'answer' to clarify the iterative process of finding relevant, though not sufficient information, reframing problems of understanding in light of new information (a new "INTU"), and progressively advancing in an area of inquiry as opposed to more textbook based, question-answer processes.

Other profitable inquiries in year 1 arose out of student curiosity relating to everyday phenomena. Textbook writers acknowledge the need to start with everyday events that students can relate to, but there are dramatic differences in how the textbook writers proceed and how students proceed. For example, D. C. Heath's *Focus on Science* text starts by having students take a deep breath and placing their hands on their chest so that they can feel their ribs expand and their diaphragm move. Silverstein and Silverstein [8] have students hold their nose and breathe through their mouths as part of their exploration of the gateways to the respiratory system. In CSILE, a child who wanted to know about the role of the diaphragm in relation to hiccuping wrote a "what I know" note. In it, she claimed to know that "everyone has their own individual laugh" – which led her to question how and why we laugh. Why are there individual differences? Reading these 'hiccuping' and 'laughing' notes inspired another child to ask "What is happening inside your body when you sneeze?" and "What is snot?"

Again, they did not continue their investigations – perhaps because of limited classroom time, but likely because the reference material was obscure, and difficult to obtain. Additionally, the failure of classroom resources to address these questions may have caused students to question the value of their inquiries. That is, if questions are not addressed in classroom textbooks, then perhaps they are not good questions to ask. Students following this line of thought begin to place a higher value on the reference material than on their own wonderings, and shift the focus of their studies accordingly. In Years 2 and 3 we attempted to deal with this issue by organizing the database around student-generated problems, and by placing greater emphasis on student questions, intuitions, and beliefs.

Another Year 1 student wondered about respiratory systems for animals besides humans. This led to the following statements and questions: "I don't know how insects breathe, but they do not have lungs." "How can there be a gas (oxygen) in a liquid (water)?" "How do whales breathe differently from humans and fish?" In attempting to answer her questions she discovered that whales have lungs like humans, whereas fish have gills. Thus whales must come out of the water to breathe whereas fish must get oxygen from the water. This led to an extended set of notes about how humans, insects, and whales breathe – with pictures of human and whale lungs, grasshopper and human trachea, etc. Then came the issue of aquatic insects.

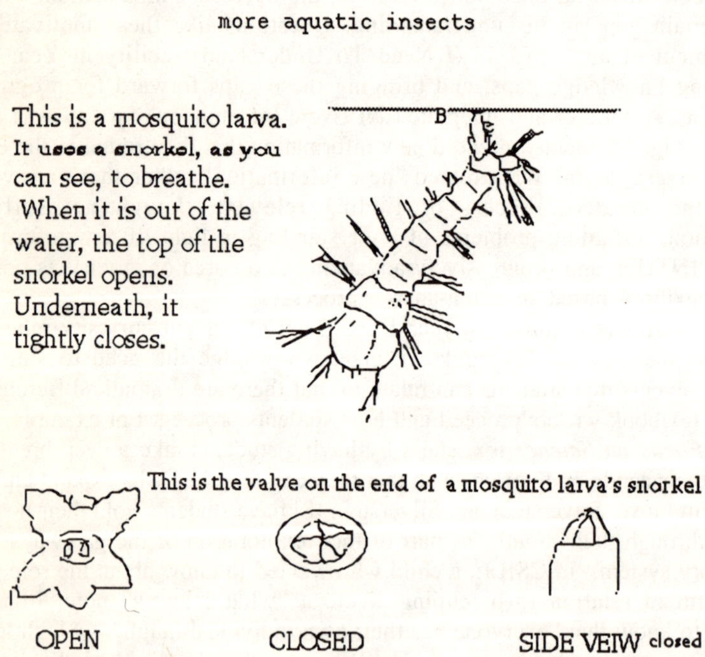

Figure 3. Breathing and aquatic insects, year 1

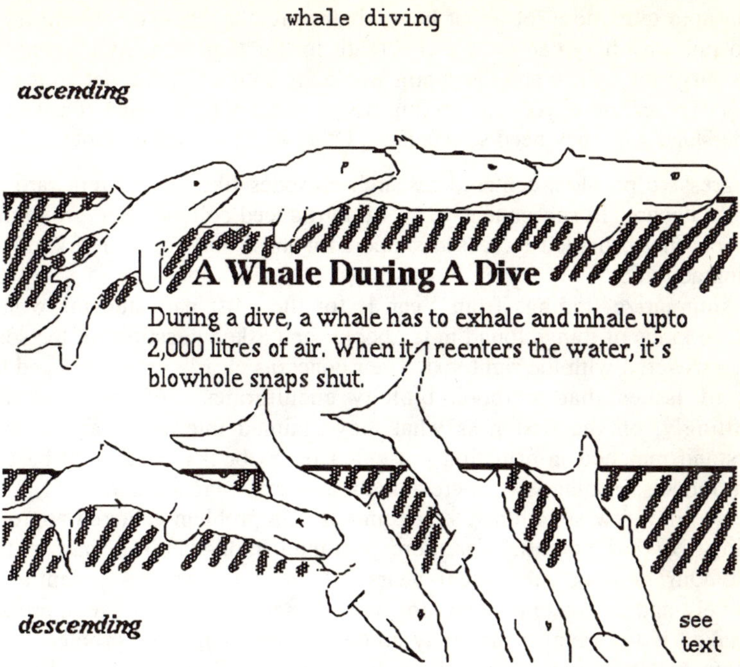

Figure 4. How Whales Breathe. Whales are mammals and they have lungs like us. That means they have to come to the surface to breathe. Whales can hold their breath much longer than people: the baleen whale can hold its breathe for up to 40 minutes, and the sperm whale for up to 75 minutes. Whales store much more of their oxygen in their muscles (about 41%,) and we store only about 13% in our muscles. We breathe through our mouths, but whales breathe through their nasal passages. During a dive, they exhale through the blowhole. Then they must take several deep breaths to restock their muscles with oxygen. They have to do all this with 2000 litres of air in a limited time. After this, the blowhole closes tightly. The heart beat slows down, and so does the blood flow, except to the heart and brain.

Figures 3 and 4 compare aquatic insects and whales. The whale picture is annotated, explaining that whales can hold their breath much longer than people – from 40 minutes (baleen whales) to 75 minutes (sperm whales) – and whales store much more oxygen in their muscles. She even notes that the whale's heart beat slows, as does blood flow, except to the heart and brain. Again, we see deepening inquiry leading to interrelationships between systems: in this case respiratory and circulatory systems. As in the prior case, the student-generated inquiry leads to interesting research topics approached from a novel perspective. Further, we begin to see signs of the kinds of supports for inquiry that are provided by students commenting on each other's notes. The student who dealt with respiration from the point of view of hiccuping and laughing read the whale note, offering this comment:

> This note tells me a lot about how whales breathe. But you should try to find out why they can hold their breath longer than humans and explain how large the baleen and the sperm whale are so that I could get an idea of what size creature you are talking about. That would make it easier to understand why they need so much air. Otherwise an excellent note.

Progressive problem-centered research episodes like these, were rare in the Year 1 database. In subsequent years we introduced collaborative group discussion tools that provided increased support for these sorts of shared knowledge-building activities.

To summarize findings from Year 1, for the most part students wondered about the kinds of things found in textbooks, and asked questions of the sort that can be answered with the right text. Their concerns clearly were mediated by the kinds of issues that textbook biology enculturates them to think about. Accordingly, on the first pass what they claimed interest in and wished to understand matched information presented in textbooks. Their text-based and mental models, insofar as we were able to tap them, were similar.

We see in a few situations the beginnings of a problem-centered approach to learning that has potential for more profound understanding than textbooks afford. Our goal in subsequent years has been to tap this potential more effectively and to create contexts for more sustained inquiry. The approach we have taken is to organize the study around student-generated problems, and to facilitate collaborative inquiry through the introduction of 'group discussion' notes. We have also provided facilities that encourage the expression of theories, and the sharing of knowledge gaps and new information. Our overall challenge is to come to terms with two serious problems highlighted by our efforts to engage students in self-directed research. First, student inquiry was modeled after textbook accounts of biological processes, which represent a topical/ structural approach to learning. In the hands of young students we find more fragmented treatment of these topics than would be the case if the contents of these texts received more systematic treatment through teacher-directed pacing of contents. Second, while there was some explanation-driven inquiry, the students did not get deeply into problems.

Years 2 and 3 Observations

Year 2 was a transitional year in which several new procedures were introduced that were consolidated in Year 3. The results in Year 2 were also transitional, showing a mixture of characteristics of the Year 1 results and those that were to emerge more strongly in Year 3. Consequently, we focus on Year 3. Our efforts in Year 1 aimed to create conditions for extended inquiry of biology text material by bringing students' mental models to the fore. We met with limited success. In subsequent years we delayed introduction of relevant text material in an effort to loosen the links between students' personally-constructed and curriculum-textbook models of learning biology. Further changes included:

Table 4. Examples of process problems: student-selected research projects, Year 3

Brain Cells: Group of 3 students
P: What kind of cells are there in the brain, and how do they differ from the other cells in the body?
P: How does the brain store information?
P: What do neuron cells look like and how do they work?
P: What are the different parts of the brain and what are they used for?

Lungs: Group of 3 students
P: How do the lungs separate the oxygen from the carbon dioxide?
P: How does smoking affect the lungs?
P: What do lungs do ?
P: How do the lungs protect themselves from the dirt we breathe, and how do they get rid of the dirt that is in the windpipe on it's way to the lungs?
P: What diseases affect the lungs, and how do you catch them?
P: How do the lungs work?

Muscles: Group of 2 students
P: Why does the muscle increase in size when it is flexed, and decrease when relaxed?
P: How do your muscles get bruised?
P: Does the size of muscles affect your strength?
P: Does the motor neuron run through, or attach, to anything else besides the muscle fibre and the motor end plate?
P: Can cartilage contract?
P: If the motor neuron is a nerve that sends electrical pulses to your muscle does it give your muscle a shock?
P: Why are you weak and tired after using your muscles for a long period of time.
P: How do your muscles get bruised?
P: How do muscles make us move?
P: How do muscle fibres contract?

The Eye: Group of 3 students
P: Where is the eye's control panel located?
P: What causes an eye to blink?
P: How does everything give or reflect light and how can a eye see that?
P: I have researched the eye and the camera and found they are very similar in many ways. I have written this note because I found that both the eye and the camera see a picture upside down. I would like to know why that happens and how.
P: How the eye sees colour.
P: How does the eye function?
P: How is the eye similar to a camera?
P: How does the message that the eye is sending get to the brain?
P: How does colour blindness occur?

(1) The students employed a direct approach to problem-centered inquiry. Each group identified a topic of interest, as in the previous year, but they had to reformulate their interests in terms of 'process problems'. Examples of process problems appear in Table 4. All subsequent investigations were organized around these problems, instead of around topics as they had been in Year 1.

(2) Students were encouraged to access mental models more directly. New thinking types were introduced that frame problems more directly in terms of personal statements of belief: my theory (MT)", and "I need to understand (INTU)." In Year 2 we noted that statements of "my theory" occurred considerably less frequently than in Year 3. In Year 3 students were encouraged to write their theories *in advance* of looking at text material. After students formulated their theories, they were encouraged to specify what they need to understand (INTU) in order to evaluate their theory. It is at this point that they were encouraged to look at texts, adding "new information" (NI) notes.

(3) A "group note" format was established. All of the students working on the same problem made entries into a common note rather than having each student write separate notes that could only be aggregated through searches of the database. The goal was to increase interaction among students and thereby support more extended inquiry.

(4) The teacher took on the role of expert learner in the class. His communications with students, both on and off CSILE, focused on the questions students were struggling with, and ways in which they could deepen their inquiry and clarify their personal theories. Some examples of his CSILE comments are shown in Table 5.

Other than these four changes, the unit on human biology was conducted much as in the previous year, starting with a brief introduction to major body systems. In Year 3, twenty-eight students formed eleven groups and selected topics as indicated in Table 4. The students used the same resource materials and continued to work in small groups and to make frequent trips between their desks and the CSILE computers. In terms of classroom organization, there were no outward differences. But the way the students approached biology changed dramatically. Consider, for example, the contrast in how students approached the topic of cells. For the most part, Year 1 notes read like material out of an encyclopedia. Research consisted of gathering facts from various sources and re-writing them in CSILE (see Table 6). The facts were not necessarily related, nor were they in any particular order. But in Year 3, students were engaged in a substantially different type of research (see Table 7). Here we see an interesting mix of probing inquiry and statements that conveyed students' naive understandings. Their ideas about biological processes were brought forward in forms that we believe could enliven and deepen biology education.

Differences between the Year 1 and Year 3 outputs are also evident at the organizational level. The topical/structural approach of Year 1 (Table 3) was completely replaced by a problem-centered approach in Year 3 (Table 6). This

Table 5. Teacher comments, Year 3

- This is very interesting. Have you any theories on why nerve cells don't grow back in some species, such as ours, but they do grow back in species such as salamanders (I think). I thought we were supposed to be the most evolved form of life! Also, do you know whether it is possible to transplant nerves the way we transplant kidneys?
- I think this is a very interesting note. I was wondering if you were going to consider how the cells differ in function? For example, do they have any special structures that enable them to communicate with other cells?
- You raise some interesting points in this note. I wonder what antigenetic proteins are? Could you explain that? What do you suppose the relationship is between antigens and antibodies?
- This is a good problem Barb. Another disease you might be interested in is emphysema. I think this disease is on the rise in our society, at least. Do you have any ideas on why?
- Lynn, I find it very interesting that red blood cells have no nucleus and are not able to reproduce. Do you have any theories on why that might be?
- Perhaps you could try to find out what a stroke is and how it affects the brain. The findings might give you some clues as to how the brain stores information.

Table 6. Cell research, Year 1

Cells

Cells are made up of atoms. Cells are the smallest common unit of life we study. There are about 10 trillion cells in your body right now. The cells in your body might look like something you might find in the sea or ocean. They might have tenticles or hair or even spikes. Cells can't be seen by the naked eye. You would need a really good microscope. When you put cells together you make tissues. All living things are made of cells. Some small things that live in the sea only have one cell. Bacteria is the smallest kind of cell. Nerve cells are the largest kind of cell. Some nerve cells are three feet long. The cell membrane gives food and oxygen for the cell to eat to make energy. Plant cells are bigger than animal cells therefore are easier to see. Plant cells and animal cells differ in many ways. Such as animal cells need oxygen to live and plant cells need carbon dioxide to live. Most plant cells can make there own sugar substance. It is made of the energy from the light and water and carbon dioxide. This substance is called photosynthesis. Each cell in your body has its own personal job. Cells do not live very long except for brain cells which you should have for life. The jelly-like stuff in the cell is called cytoplasm. The nucleus controls the cells reproduction. The blood brings digested food to the cell. Cell is a Latin word that means 'a little room'. The cell became known after they invented the microscope because you couldn't see the cell with your naked eye. When carbon dioxide gets near the cell, the cell will not let it in. The organelles are like our organs.

Table 7. Cell research, Year 3

Problem: How does the eye function?(BN)

My Theory: I think that the muscle in the eye can let it go in any direction it wants so that it can see anything without turning the head around. Also I think that there is a special part of it that can allow the brain to see the picture rapidly and still have a good look at it. (BN)

I Need To Understand: What the muscles of the eye do. (BN)

New Information: The eye is in the nervous system. Special senses such as the eye, are called the sense organs. There are six muscles attached to the eyeball that can move it in any direction the brain wants it to. The muscles of the iris can make the pupil increase or decrease its size. (BN)

My Theory: I think that there is almost a kind of filter behind the eye where the picture goes through transforming it into waves which the brain understands. (NT)

I Need To Understand: How the eye sends pictures to the brain. (BN)

New Information: I have gotten some new information on the eye. I have read that every object gives or reflects light. The eye reads that reflection and sends its message to the brain where it can be identified. (BP)

New Information: There is a black panel behind the eye which absorbs the light and makes sure none can get out. (NT)

New Information: An object that the eye is focusing on gives a ray of light and goes by the cornea, the aqueous humor, the pupil, the lens, the vitreous humor and then it reaches the retina. When the light hits the cells of the retina they react the nerve cells and then the optic nerve brings the message to the visual cortex hitting part of the brain which reads the message. (BN)

I Need To Understand: Why the pupil increases and decreases. (BN)

My Theory: As I already found out that the muscles of the iris can increase or decrease the pupil, I think that when there is a strong source of light the iris tries to protect it self from the light and uses its muscles to decrease the pupil and when it is doing so the eyelid slightly closes. (BN)

New Information: Everything the eye sees is really light reflecting off the object. (NT)

I Need To Understand: How the parts of the eye help get to the message to the brain. (BN)

New Information: When an object gives the light it hits the curved surface of the cornea. After leaving it, the iris increases or decreases the pupil and it goes by. If there is a strong source of light the pupil will decrease to shut off light that is not needed. After the light rays pass the pupil it passes the curved lens then it passes the vitreous humor and hits the retina where the light rays meet. If the light rays meet on the retina the image is clear and if it goes inside or in front, the image will be blurred. Then the light passes the retina and goes through the nerve cells to get to the rods and cones. The cones are in about the center of the retina, they also do most of the seeing during the day. They give clear images with a lot of detail. The image is brought by the optic nerve and is read by the brain. (BN)

---

new focus on student-generated inquiry allowed us to better observe student situation models and the evolution of these models during the learning process. We made the following observations.

Student-generated theories: A context for figuring out how things really work

In Year 1, students rarely expressed their theories about biological processes. But in Year 3 their theories provided a context from which they could begin to figure out how things work. Theory construction was encouraged even in situations in which the student knew little, or nothing, about the underlying processes. Consequently, many student theories were simplistic, or contained misconceptions. For instance, the following child began her investigation of cells by proposing that each limb has a specialized cell type:

My Theory: I think cells tell all the parts of the body what they are supposed to do. So an ear cell tells an ear what to do, an arm cell tells an arm what to do and leg cells tell a leg what to do.

Initially, we were afraid that misconceptions such as these might be propagated among students. But a study of classroom beliefs determined that this was not the case [1]. Students seemed to recognize that entries labeled as "my theory" were attempts by their peers at understanding biological processes, and not necessarily scientifically accepted explanations. Instead, theories provided a foundation upon which the learner could build an investigation, and begin a discussion with peers. Bringing the situation model to the fore led to new ideas and questions, as can be seen in the following example:

My Theory: I think that antigens and antibodies are the same. I think that antibodies kill germs in the body. Maybe when these germs (if they are the same germs) enter the body again the antibodies kill them before the disease develops. That is my theory on why you usually only have chicken pox once. I wonder why you have colds more than once then! I think that the antibodies in your body do not always destroy some diseases or viruses because they do not recognize them as a harmful substance. That is my theory on why some diseases, such as AIDS cannot be naturally cured.

Later, this student discovered that antigens and antibodies were *not* the same, and also learned how vaccinations work with the immune system to fight disease.

Student-generated questions:
A context for identifying gaps in understanding

As follow-up work to their statements of belief, or their analysis of text material, students in Year 3 formulated "I Need To Understand" statements. Examples of these entries are shown below:

I Need To Understand: How when you are sleeping, do you keep breathing and not stop and die?

I Need To Understand: How muscles are kept from falling off the skeleton.

I Need To Understand: How does the nerve cell send messages?

I Need To Understand: When the veins bring the blood containing carbon dioxide and other waste products back to the right chamber of the heart how does it get rid of it?

I Need To Understand: *How* does the blood take the oxygen from the lungs? How does it deliver it to organs? Why does our body need oxygen?

I Need To Understand: Why do red blood cells have no nucleus and do not reproduce?

I Need To Understand: How does the long-term memory store the information?

These inquiries subsequently led to deeper investigations of the subject matter. Some problems that began as an investigation of a single organ (e.g., "How do the lungs work?") produced sub-questions that broadened the investigation into a study of an entire biological system, e.g., the circulatory system. Overall, students demonstrated a remarkable capacity for figuring out what they needed to know to advance their understanding. They seem to have greater insight into their mental models than curriculum writers have [5]. Almost all INTU statements inspired further, deeper discussion although few were completely resolved. But in contrast, many Year 1 questions were not addressed at all.

Student Research: Finding information to fill gaps in understanding

Year 3 students found relevant, though seldom complete information regarding the issues they raised. Most of their information was found in reference books from the library, CD-ROM materials, or the encyclopedia, and then entered into

CSILE as "New Information" entries. Two examples are shown below. The first example is taken from an investigation that began with the problem, "How do the lungs separate the oxygen from the carbon dioxide?" The second example is from the problem, "How do muscles make us move?"

New Information: When the oxygen comes into your lungs from the bronchial tubes, it passes through a wall of something called alveoli. An alveoli looks like a bunch of balloons mashed together. Inside the alveoli are some tiny blood vessels, called capillaries. After the oxygen passes through the alveoli and capillary walls, it gets sucked into the red blood cells. The red blood cells need more oxygen, and they have been traveling around your body collecting carbon dioxide. By the time the red blood cells get back to your lungs, all their oxygen has been turned into carbon dioxide. They have a lot of carbon dioxide to get rid of, and a lot of oxygen to bring in.

New Information: The steps of a muscle contracting in an arm: (1) your brain sends an electrical message through the axon of the motor neuron, (2) the motor neuron stimulates the muscle fibre, (3) this makes the muscle fibers enlarge, so the muscle can contract, (4) now the arm can move. It may seem like a long process, because the way I'm explaining the steps one by one, but this all happens in a split second.

As in Year 1, these students extracted information from various resource materials, and brought the knowledge back to CSILE. But the Year 3 students were guided more by their own developing model, and less by the organization of the textbook. In Year 1, students extracted information somewhat indiscriminately from the texts available to them. In Year 3, they tended to sift out those materials that were relevant to the problems they were trying to solve. Personal knowledge needs guided their research efforts. Unfortunately, inadequacies in the resource materials frequently blocked students from moving deeper into problems.

Revised Explanations

Year 3 students almost always revised their explanations before the termination of the investigation. Curiously, they rarely used "My Theory" to express their revised explanation, preferring to use "New Information" instead. The one exception to this occurred during an investigation of color blindness. The initial theory is shown below:

My Theory: I think that it happens by the lens getting damaged and a certain quality of it being lost. This quality, I think, enables the eye to see things in colour.

After some further questions and the acquisition of additional information, the student produced the following revision:

My Theory: When the light rays reach the retina, they pass through nerve cells to reach the rods and cones. The cones lie mainly in the macula, a small area near the center of the retina. They are less sensitive then the rods and respond only to pretty strong light. They are responsible for colour vision. The rods cannot see colour and I think that colour blindness is caused when the cones are damaged and only the rods are left which can't see colour.

Revised explanations are useful because they make knowledge advances apparent to the student, as well as to other readers of the database. The tendency for Year 3 students to express these revisions in "New Information" statements was initially surprising to us, but upon reflection, a logical product of an explanation-centered programme. Rather than use "New Information" to simply list facts gathered from their research investigations (as they had done in Year 1), the students were also attempting to synthesize those facts into a coherent, logical, explanation. "New Information" entries like those in part (c), and the revised "My Theory" shown above are examples of students attempting to build increasingly more explanatory models of biological processes during their investigations.

Discussion

Why should such small changes in procedure and shifts in emphasis produce such a change in the way students used texts in learning? It appears that getting students to venture their own 'theories' in advance of getting into texts is crucial. Finding an adequate theory to explain phenomena then becomes the guiding purpose, resulting in a more goal-directed and constructive approach to text information. Our own 'theory' about this phenomenon is as follows: As suggested at the beginning of this paper, constructive learning from text requires bringing one's own mental model to the text [4, 9]. But what if you do not have a mental model to bring? We cannot be sure whether this is the case with the students we have been working with or whether they have mental models that are not brought to mind. The result is the same in either case: If they start reading texts, with no mental model of their own brought into the process, they are inclined to read passively and to be dominated by the text model (which is not a theoretical model that the text may be referring to but the topic-detail model of the text itself). What the Year 3 procedure does, we believe, is get students to construct provisional mental models, which they can then bring to texts. A testable but as yet untested hypothesis is that this results in new information being organized according to the mental model rather than according to the text structure. In any event, it leads to more active, constructive processing of text information, with greater relevance to problems of understanding.

We find that students are initially reluctant to present provisional mental models. They know that they are just making something up and that they lack essential information. One of the virtues of getting them to do so is that it sets them up for the next step, INTU ("I need to understand"), in which they try to identify the information they actually need to produce a more adequate theory. There are probably other ways of getting students to develop or activate mental models in advance of going into texts. Biology experiments, observations, and so on might provide stimulus and some basis for theorizing. But if our interpretation is correct, 'hands-on' activities will only be effective if they are carried forward to the point where students produce their own provisional explanatory models.

In follow-up experiments we plan two classes of activity: to link students with biology experts and to take advantage of the pockets of local expertise that arise from student specializations. With regard to the former, we expect to demonstrate that CSILE students need less access to content specialists than is typically assumed, and when they have expert access, they make better use of it. Procedures used in Year 3 foster local specializations, because individual students do a great deal of thinking about specific issues such as blood clotting, breathing, etc. The database that they construct via CSILE allows students to begin viewing their work in the aggregate. We believe that local pockets of understanding will expand in fascinating ways, as students who have studied a process deeply make contact with others who have studied related but different phenomena deeply. Overall, we plan to continue creating profitable contexts for student inquiry. Using new knowledge media and new knowledge processes, we hope to expand student understanding of biology.

Acknowledgements

The authors wish to acknowledge the generous support of the James S. McDonnell Foundation; the Ontario Ministry of Education under its block transfer grant to the Ontario Institute for Studies in Education; the Ontario Ministry of Colleges and Universities through the University Research Incentive Fund; and the Social Sciences and Humanities Research Council of Canada. We are indebted to the Grade 5–6 students at Huron Public School who contributed their time and talents to this project. We are also grateful to Peter Rowley, Earl Fogel, Wayne Liu, Russell Owen, and Alan Rosenthal who made the software modifications necessary for this study. Finally, we'd like to acknowledge the efforts of the entire CSILE team, without whose contributions the work reported here would not have been possible.

References

1. Burtis, P. J., Chan, C., Hewitt, J., Scardamalia, M., & Bereiter, C.: Misconceptions in a student-guided research project. Paper presented at the Annual Meeting of the American Educational Research Association, Atlanta, GA. April, 1993.
2. Cohen, A.: Using CSILE in a progressive discourse in physics. Paper presented at the annual meeting of the American Educational Research Association (AERA), San Francisco, CA. (1992).

3. Heap, J. L.: Discourse in the production of classroom knowledge: Reading lessons. Curriculum Inquiry, 15(3), 245-280, 1985.
4. Kintsch, W., & van Dijk, T. A.: Toward a model of text comprehension and production. Psychological Review, *85*, 363-394, 1978.
5. Leinhardt, G.: Towards understanding instructional explanations (LRDC Tech. Rep. No. CLIP-90-03.) Pittsburgh: University of Pittsburgh, Learning Research and Development Center (1990).
6. Roth, K. J.: Conceptual understanding in higher level thinking in the elementary science curriculum: Three perspectives (Elementary Subjects Centre Research Series #12). East Lansing, MI: Michigan State University, Institute for Research on Teaching, 1990.
7. Scardamalia, M. & Bereiter, C.: Higher levels of agency for children in knowledge building: A challenge for the design of new knowledge media. Journal of the Learning Sciences, 1(1), 37-68, 1991.
8. Silverstein & Silverstein: The respiratory system: How living creatures breathe. NJ: Prentice Hall 1969.
9. van Dijk, T. A. & Kintsch, W.: Strategies of discourse comprehension. New York: Academic Press, 1983.

Darwinian and Lamarckian Models Used by Students and Their Representation

Maria Pilar Jimenez Aleixandre

Department of Didactica das Ciencias Experimentals
Universidade de Santiago de Compostela, Santiago, Spain

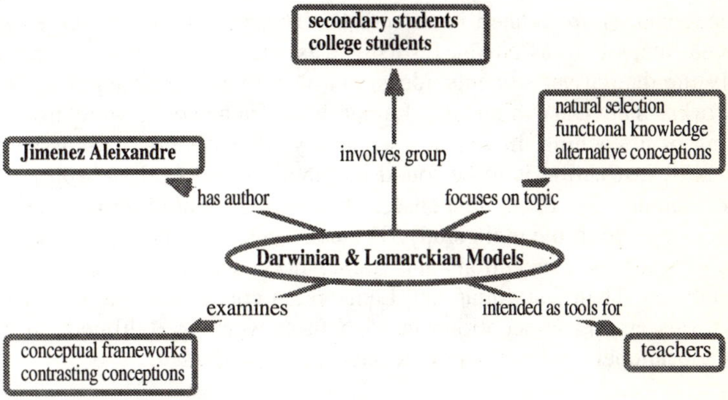

Abstract. As a part of a larger study on the learning of natural selection by secondary school and university students, an attempt was made to identify conceptions that belong to the Darwinian model, taught in school science, and the Lamarckian model, the alternative view most widely used by students. This paper focuses on the different conceptions that make part of the frameworks used for explaining biological change, and the way they are related or structured. These relations are represented here as conceptual maps, intended to be used as tools in instruction and teacher training.

Keywords. Knowledge representation, Natural selection, Conceptual maps, Students' ideas, Functional knowledge

Introduction: Students' Knowledge About Biological Change

One of the goals of science instruction is to promote the reconstruction by pupils of the current school-science view. To do this it is useful to know students' conceptions on the topic, as well as their relationships. As part of a larger study [14] on the learning of natural selection, following the conceptual change model

Introduction: Students' Knowledge About Biological Change

One of the goals of science instruction is to promote the reconstruction by pupils of the current school-science view. To do this it is useful to know students' conceptions on the topic, as well as their relationships. As part of a larger study [14] on the learning of natural selection, following the conceptual change model [13], an attempt was made to identify conceptions that belong to the Darwinian model, taught in school science, and to the Lamarckian model, the alternative view most widely used by pupils.

In this paper I will focus on the different conceptions that make part of what I will call the 'frameworks' used for explaining biological change and the way these conceptions are related or structured. These relations are represented as conceptual maps that, as Novak [20] suggests, are useful ways for representing not only the data about students' ideas, but also about the changes in ideas and frameworks. The Darwinian and Lamarckian frameworks were used in the designing of a teaching module for secondary school pupils. Discussion of the outcomes of instruction is to be found in another paper [15], as well as a more detailed account of students' responses. The method we used for categorizing the responses is given in the main study. The next section of the paper discusses the notions of students' conceptions and frameworks, and the way the conceptions were analyzed. Then Darwinian and Lamarckian conceptions are described, and the relations among conceptions in each framework are outlined. In the last section educational implications of the analysis are discussed.

Students' Conceptions and Frameworks

There is an ongoing discussion concerning whether frameworks exist or not. Some authors, like Clough & Driver [5], think that the fact that some of the alternative ideas about science phenomena that are used by different people in a variety of contexts and with a certain degree of consistency supports the existence of frameworks. However, it has been argued [18] that students' sets of ideas lack coherence. They may not represent a real model or 'theory' the student was holding, but rather a random answer to some question whose solution the student did not know.

In my opinion, a more interesting idea of conceptual frameworks is it provides a useful tool to analyze students' conceptions, organize them, and use them in designing instruction. I believe that conceptual frameworks exist to some extent and are widely shared because we have found the same kinds of explanations among a sample including 773 students, with explanations and ideas similar to those found in studies in other countries [8, 4, 3]. A different question is the way in which students use their 'models', which may lack the consistency and the kind of epistemological commitments [12] that would be expected of scientists when using theirs.

In this paper I define 'conception' as an idea with a limited scope, a part of an explanation about a topic such as biological change, but not complete in itself. I will define 'framework' as the structure formed by a number of these conceptions with particular relations among them. It may be noted that the students use these frameworks as 'models' in their interpretations.

As these conceptions and models attempt to represent the ideas of a large number of students, and not of a particular one, they constitute some kind of ideal representations, and not verbatim wording. West et al [22] have discussed the contradiction in trying to represent a group summary of idiosyncratic cognition, and, as they say, the representation will focus on the ideas and relations shared among learners - in this case the ones in our sample.

The method used to study the ideas about natural selection was to propose some questions that attempted to assess procedural (in the sense of functional, that is one which can be transferred to a different context) knowledge, by asking the students to interpret novel situations. In none of the questions did the words 'evolution', 'adaptation' or 'natural selection' appear. The answers were then analyzed and coded, using a number of categories [16]. Only conceptions in two of those categories will be discussed here: Darwinian, the current school science model, and Lamarckian, the most widely used by pupils. The representation has two parts which correspond to the two components of cognitive structure [22], knowledge bits, in our case propositions, that are discussed in Sect. 2, and how the knowledge is organized, discussed in Sect. 3.

Darwinian and Lamarckian Conceptions

What here are termed 'Darwinian' conceptions belong to the neo-Darwinian synthesis and are more precisely what is taught in school science. That may not match exactly the academic view, being a simplified version. That is, it does not correspond wholly to historic ideas, first because it includes aspects relating to DNA and mutation, not known in Darwin's time, and second because as Beddall [2] and Gruber [10] have shown, Darwin believed in some mechanisms like use and disuse that contradict the current genetic view.

What here are termed 'Lamarckian' conceptions are the explanations of biological change as effects of the environment, resulting in changes in individuals and inheritance of those acquired traits; but also included among them are ideas about changes in DNA due to environmental factors, and some finalist and intentional ideas that appear in close association with Lamarckian ones in students' responses, independently of them belonging to the thinking of Jean-Baptiste Lamarck, to whom we owe the first outline of an evolution theory.

The two sets of conceptions and the models are presented here as clearly distinct, and in opposition to one another. But it is worth noting that in a number of students' explanations, conceptions of both frameworks get mixed, showing that to them these ideas are compatible. Thus, the main focus of the classroom study [15] was to evidence the existence of two incompatible models for explaining biological change, and to develop students' awareness of their own ideas by making them switch consciously from one to the other.

In Table 1 the conceptions are listed, describing each idea in accordance with school science labeled D1, D2, etc., for Darwinian ideas and L1, L2, etc., for the corresponding alternative Lamarckian ideas. Conceptions D11 to D 15 and L11 to L15 can be considered auxiliary conceptions, and due to the lack of space, will not be discussed further.

The extent of each conception will be analyzed below, with an indication of whether they are ideas explicitly used by students in our sample, or in other studies, or if they are implicit in their interpretations.

D1 Intra-species variation. This idea, together with D3 – excessive offspring, and D5 – differential survival, form the core of Darwinian theory. In the students' framework, this idea is explicit in a number of answers that establish a distinction between two types of individuals, e.g., resistant and non-resistant, in the population.

L1 Species uniformity. The corresponding idea in the Lamarckian framework is usually not formulated, but lies in answers that interpret changes as occurring in individuals. If students perceive only one explanation for the transformation of individuals, it may be because they don't perceive that some could already possess a given trait before change in the environment. In my opinion this idea is related to a biased view of biological heredity as a mechanism that preserves – from one generation to the next – similarities, and not as a mechanism that preserves both *similarities* and *differences*. This, in turn, is linked to a view of individual genetics and not to population genetics.

D2 Fortuitous mutation. This is one of the conceptions that has its origin in genetics theory, and fills a blank in Darwin's model with respect to the origin of variation. The point is that mutations caused by environmental mutagenic agents do not possess adaptive value with respect the specific agent (for example, mutations due to radiation in atomic explosions do not confer protection against radiation). This idea appears explicitly in students' answers in the higher grade levels; they also mention recombination as a source of variation.

L2 Induced mutation. Appears explicitly in a number of responses in our study and in Brumby (1979).

D3 Excess offspring. This idea as well as L3 lacks an explicit formulation, being implicitly linked to D4 (intra-species competition) as a requisite to explain differential survival and change in population, because if all offspring were to survive, proportions between different traits would stay the same.

L3 All offspring survive. Direct experience of students on this issue with human beings and domestic offspring may lead them to believe that survival of all offspring is normal and not an abnormal condition, a conception that is implicit in the Lamarckian framework.

D4 Intra-species competition. This idea has suffered much erroneous interpretation in history, because the 'struggle for life' has been understood at times as an actual physical struggle. In students' answers, explicit references to competition appear - or to the absence of it, when for instance non-resistant individuals are eliminated.

L4 Physical struggle. Competition appears in the Lamarckian framework primarily as physical struggles between different species, an aspect noted also by Deadman & Kelly [8]. As an extension of this image of physical struggle, students often identify 'the fittest' with 'the strongest', mentioned too by Hallden [11] and Bishop & Anderson [3]. This view is associated with Darwinian as well as Lamarckian answers.

D5 Differential survival. In my opinion, differential survival is the key idea in the understanding of natural selection. In students responses, the idea of survival of more offspring from better adapted individuals is explicit when they are using the Darwinian framework.

L5 Slightly better adapted. The idea about individuals in each generation being better adapted than those in the previous generation is often explicit, as Brumby [4] also mentioned. Students think of changes in individuals rather than changes in the population, as reflected in L6. The wordings are often '*each time they are more resistant*' or '*each time they have shorter tails*'.

D6 Population changes. The understanding of evolution as a change in the proportions of different genotypes in the population is closely related to D5. In 'Darwinian' answers, this idea rarely appears explicitly, being implicit in D5.

L6 Individuals change. The transformations experienced by individuals are explicit in most cases. This idea is related to L5 and also to L15 (time scale), and to the difficulties students have in conceptualizing processes are imperceptible that in the scale of human life.

D7 Pre-adaptation. Problems in learning natural selection are related to the concept of adaptation. Pre-adaptation means that different traits exist in a population that may be advantageous, depending on the differing environments. In students' responses this conception is seldom explicit.

L7 Post-adaptation. Adaptation is viewed as an *a posteriori* mechanism. Students frequently use '*they adapt*' with an active meaning, interchangeable with '*they get used to*', '*they adjust*', etc. This has been mentioned by other authors [8]. A consequence of this idea is that *ad hoc* advantageous traits appear under different circumstances.

D8 Acquired traits not inherited. Darwin didn't completely reject the inheritance of acquired traits. Nevertheless, in the current Darwinian model, this idea is not compatible with D2 (that is, DNA is not modified as an adaptive response to environmental mutagens). In students' answers this conception appears in a few occasions, in the upper grade levels.

L8 Acquired traits inherited. This idea appears both explicitly and sometimes implicitly as in L2. Other studies mention it [8, 5, 17].

D9 Useless organs exist. As in D8, Darwin didn't reject the use and disuse law, but it is nonetheless incompatible with current genetics theory. Recognition of this idea appeared in student answers on a few occasions.

L9 Use and disuse law. This conception appears explicitly, particularly in the answers to the 'mice' question.

D10 Genetic resistance. This conception is a particular instance that is included because of the occurrence of L10. In the answers, the idea of innate

Table 1. Conceptions about natural selection

Darwinian	Lamarckian
D1 Intra-species variation. There is individual variation in some inherited traits in a single species.	**L1 Species uniformity**. Individuals of a single species are all alike.
D2 Fortuitous mutations. Mutations are due to chance or to environmental agents, but without being adaptive responses to these agents.	**L2 Induced mutations**. Mutations are induced by the environment as adaptive responses to changes in that environment.
D3 Excessive offspring. Organisms produce many more offspring than can survive.	**L3 All offspring survive**. The number of surviving offspring is similar to the number of offspring produced.
D4 Intra-species competition. There is competition among members of one species for place, food, etc. which resolves in favor of those best suited to the current environment.	**L4 Physical struggle**. There is physical fighting among one species – or among different species – and the strongest win.
D5 Differential survival. Individuals with a given trait (best adapted) will contribute more offspring to the next generation.	**L5 Slightly better adaptation**. Nearly all individuals become slightly better adapted in each successive generation. Those not adapted will perish.
D6 Populations change. Populations change through increasing or decreasing the proportion of individuals with one or several traits.	**L6 Individuals change**. Populations change through changes in individuals, wherein an individual loses or gains a trait.
D7 Pre-adaptation. A change in the environment may select a trait by favoring those individuals who have that trait.	**L7 Post-adaptation**. A change in the environment induces changes in individuals, such that they become better suited to that environment.
D8 Acquired traits not inherited. Only traits specified by alleles in the genomes of the parents' sex cells are inherited.	**L8 Acquired traits are inherited.** Advantageous traits (such as physical fitness or short tails) that are acquired by parents can be transmitted to offspring.
D9 Useless organs exist. An organ may exist even if it is not used unless eliminated by a mutation.	**L9 Use and disuse law**. When an organ is not used it undergoes atrophy and finally disappears.
D10 Genetic resistance. An insect or bacterium does not acquire resistance to an insecticide or antibiotic upon exposure to it (typically, some individuals have resistance by chance mutation).	**L10 Acquired 'immunity'**. An insect or bacterium acquires resistance to an insecticide or antibiotic upon first exposure to it.
D11 Adaptation independent of will. Living things express adaptive traits or don't, independently of their will.	**L11 Intentional adaptation**. Living things acquire adaptations as an act of will in order to survive, often in response to environmental pressures.

Darwinian	Lamarckian
D12 No final causes. A species can become extinct even if it plays an essential role in the ecosystem.	**L12 Final causes**. Species with given roles in the ecosystem do not become extinct.
D13 Humans comprise one species. A species can become extinct even if it is useful to humans (or even if it is the human species).	**L13 Anthropomorphism.** A species survives or disappears depending on its utility to humans.
D14 Probabilism. Proportions of traits among offspring are probabilistic.	**L14 Determinism**. Proportions of traits among offspring are given.
D15 Geological time scale. Time scale in evolution is measured in eons.	**L15 Human time scale**. Evolutionary changes occur in a few generations.

resistance appears some times explicitly and other times implicitly in responses that indicate that some individuals die and others do not when exposed to an insecticide or antibiotic.

L10 Acquired 'immunity'. In the Lamarckian framework there is a confusion between resistance and immunity, related to an identification of resistance as an active defense against certain environmental factors and not as an innate feature. It often appears explicitly – see also Brumby [4] and Hallden [11].

How the Conceptions Are Organized

Figures 1 and 2 are attempts to represent how the conceptions are organized. The complexity of the topic forces the use of elements in the map of 'conceptions' that are, in turn, complex ideas – as can be seen in Table 1 – and not simple concepts as in other topics. Two necessary steps when constructing the maps were: (a) to establish the number of relations between one conception and the other; thus in the Darwinian map, conceptions such as biological heredity or differential survival have a more complex set of relations than others; (b) to name the relations. According to Fisher [9] classifications such as those presented here are of a logical nature. A question raised by Ausubel et al. [1] is the extent to which a correspondence does exist between a *logical* conceptual framework, that is, the relations among ideas from the discipline perspective, and a *psychological* framework, or the relations among the ideas in the students' minds. In our case this only affects the Darwinian model, because in the Lamarckian we are dealing mainly with students' ideas.

Some conceptions constitute what we may call the core of the framework, while others can be viewed as consequences or implications of the first. From a logical viewpoint – say from the perspective of the scientific explanatory model – these conceptions are not necessary in the framework; they are additional or auxiliary conceptions that may or may not be present, without the interpretation being substantially changed.

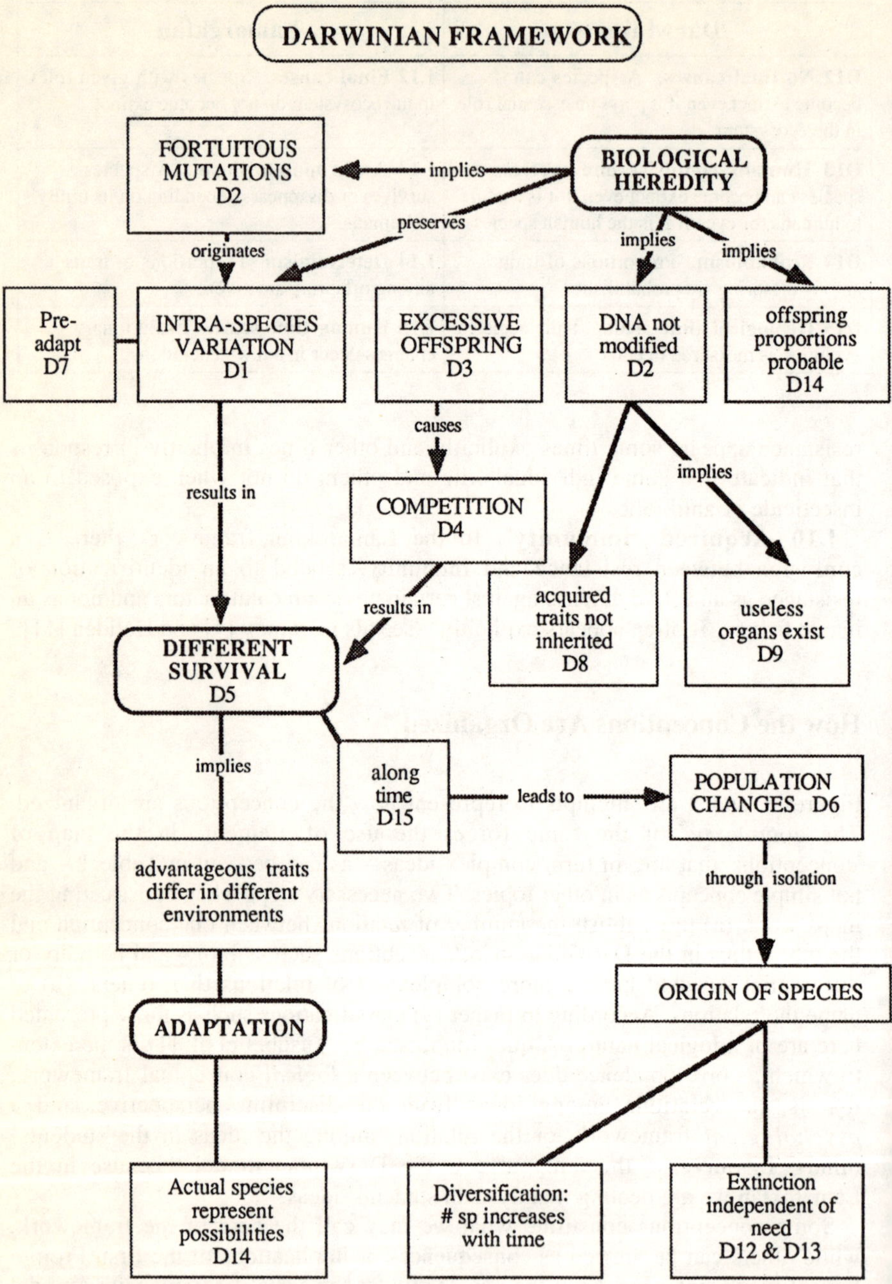

Figure 1. Darwinian Framework

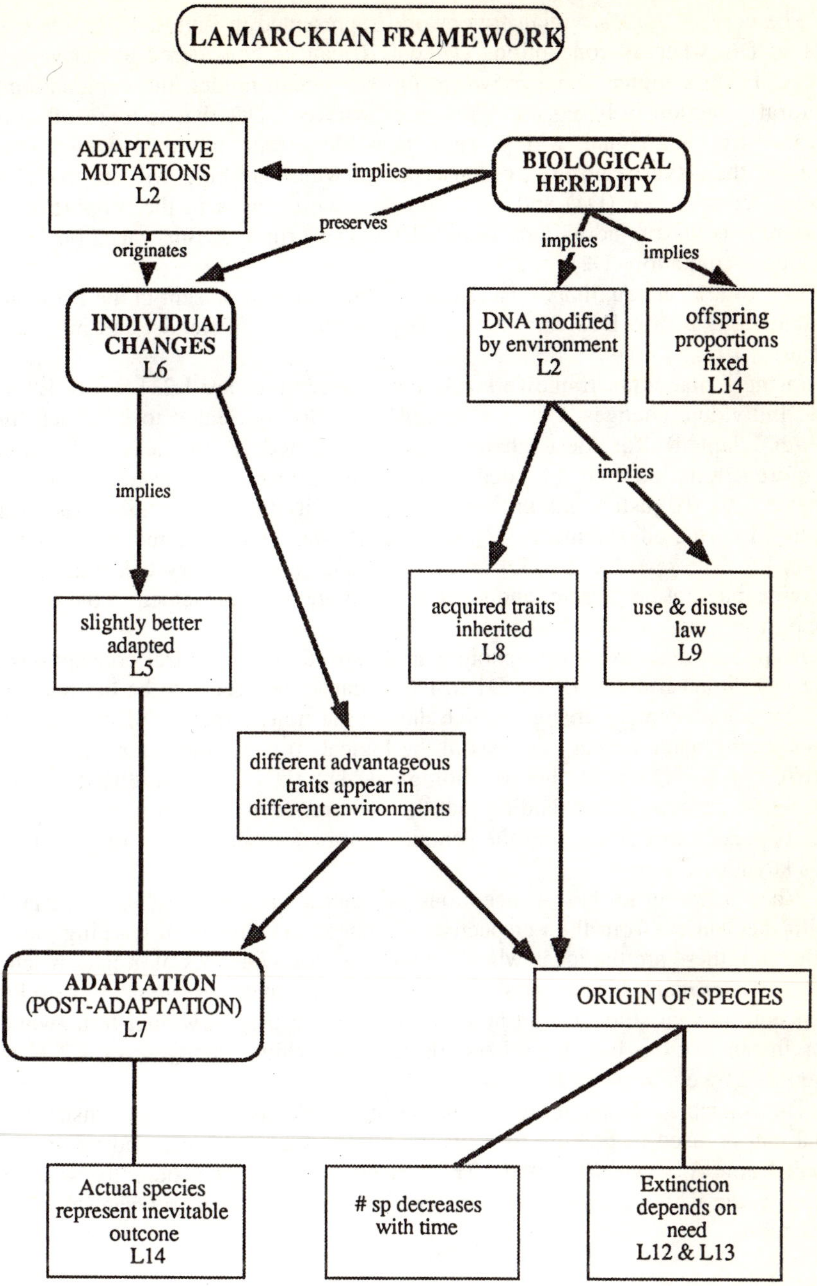

Figure 2. Lamarckian Framework

The core of the Darwinian framework, represented in Figure 1, are the ideas D1 to D6, whereas conceptions D7 to D10 can be considered to derive from these. From a logical perspective, in the Darwinian model, the explanation of natural selection is based on *differential survival,* D5, that is more offspring survive from individuals with given traits and less from individuals which do not present them (variation D1), because not all which are born will survive (D3), due to competition (D4) and, as a consequence changes in the proportions of different types of individuals occur (D6). The origin of this variation lies in fortuitous mutation (D2).

D11 to D15 are auxiliary conceptions and may not be present in the Darwinian framework; D11 to D13 can be considered refutations of finalist and intentional ideas.

In the Lamarckian framework, the core conceptions are L2, L5, L6, L8 and L9. Individual changes, L6, is a central idea, closely related to L5 (each time better adapted); for these changes to be reflected in the next generations, acquired traits must be inherited (L8) which, in turn, is connected to use and disuse (L9). To justify this in terms of the heredity theory, mutations should be induced by the environment (L2). L7 and L10 can be considered consequences from L2. The rest, L1, L3, L4 and L11 to L15, are auxiliary conceptions that may or may not be present, and some of them are not represented in the concept map in Figure 2.

It appears that, from a psychological perspective, the central conceptions in the Darwinian framework are D1 to D6, because they seem to be the ones that determine conceptual change, which entails the transformation of the rest. The point being made is that, whereas in the logical structure the central idea is D5 (different survival), in the psychological, D2 (DNA not modified) is also essential, because understanding that the occurrence of a given feature depends on its presence or absence in the genetic information, and not for other reasons, is a key idea.

When referring to these conceptions as 'central', as other studies [19, 21] did with mechanics, from the perspective of conceptual change in learning natural selection, these are the ideas whose change has consequences of greater reach in the *restructuring* of the whole. In other words, there are some Lamarckian conceptions that students can hold while adopting the Darwinian framework – for instance L4 – but not others like L6 (individual changes) or L2 (DNA modified by environment).

The conceptual maps in Figs. 1 and 2 should be considered ideal constructions that can be used in helping to visualize relations among ideas and to compare each framework with the other. The concept of ‘biological heredity’ does not refer to the whole of genetic theory, but only to those ideas relevant to natural selection.

Conclusions and Some Implications for Teaching

In the conceptual structure, ideas about evolution are closely related to genetics knowledge because, first, as detailed in the main study, the decline in alternative ideas matches the increase in genetics knowledge, even if the students have not received further instruction. Second, students argue by recourse to heredity as an explanation. This happens with both the ones using the Darwinian model as well as with those who use the Lamarckian one.

When analyzing the responses of the older students (university biology students), it was found that a higher number of those who use the Darwinian framework were justifying them in terms of genetics theory, for instance the information flow. But what is a cause for concern is that a higher number of those using the Lamarckian framework also would refer to 'changes in DNA' or 'crossing-over' to justify their answers. That, in my opinion, is an example of rote learning, in which technical jargon and bits of information are learned, but do not find an appropriate structure to fit into.

Another issue is that differences between school science and alternative ideas are more pronounced in relation to procedural and functional knowledge. Students, even those giving Lamarckian answers, are able to use a lot of terminology related to evolution, but they show difficulties in applying the Darwinian model to novel situations, to transfer the knowledge to problems set in different contexts.

It is not possible in these pages to detail the use of the conceptual maps in instruction. The ones presented here were used in teacher training and with secondary school students. Some of the activities for teachers consisted of:

- Asking them to list propositions about natural selection, and constructing a map with them. We observed that some key propositions, e.g., 'excessive offspring', were lacking.
- Providing the list of conceptions and asking them to organize the map and name the relations. Some of the problems found related to 'adaptation'.

The Lamarckian map was also used to discuss with teachers the alternative ideas pupils hold.

We employed simplified versions of the list of propositions with students and asked them to construct their own maps at different moments in the unit, and to compare them at the end. We found this to be a powerful tool in enhancing their metacognition, in making them aware about the models they hold, and how they are using them.

In summary, we believe that these maps – and others similar in different topics – constitute a resource for teacher's previous work. It helps them to become aware, in a systematic way, of the difference between students' ideas and school science – and as the main study [14] evidences, teachers usually don't pay much heed to this question – as well as to help students to 'learn how to learn' about natural selection.

Acknowledgments

To Mike Beeth for his helpful comments on an earlier draft.

References

1. Ausubel, D., Novak, J. & Hanesian, H.: Educational Psychology: cognitive view, Holt, Rinehart and Winston 1979.
2. Beddall, B.: Wallace, Darwin and Natural Selection: a study in the development of ideas and attitudes. J. of History of Biology, 1, 261-323, 1968.
3. Bishop, B. & Anderson, C.: Students Conceptions of Natural Selection and its role in evolution. Journal of Research in Science Teaching, 27 (5), 415-427, 1990.
4. Brumby, M.: Students' perceptions and learning styles associated with the concept of Evolution by Natural Selection. Ph. D. Thesis, Surrey, 1979.
5. Clough, E.E. & Driver, R.: A study of consistency in the use of students' conceptual frameworks across different task context. Science Education, 70 (4), 473-496, 1986.
6. Clough, E.E. & Wood-Robinson, C.: How secondary students interpret instances of biological adaptation. J. Biological Educ., 19 (2), 125-130, 1985.
7. Darwin, C. & Wallace, A.: 1859 On the tendency of species to form varieties and on the perpetuation of varieties and species by natural means of Selection. J. of the Linnean Society of London (Zoology), 3, 45-62 in Harris Evolution, Genesis and Revelation, State Univ. of NY Press 1981.
8. Deadman, J. A. & Kelly, P. J.: What do secondary school boys understand about evolution and heredity before they are taught the topics? Journal of Biological Education, 12 (1), 7-15, 1978.
9. Fisher, K.: Semantic Networking: the new kid on the block. Journal of Research in Science Teaching, 27(10), 1001-1018, 1990.
10. Gruber, H.: Darwin on man. A Psychological study of scientific creativity. The University of Chicago Press 1981.
11. Hallden, O.: The Evolution of species: pupil perspectives and school perspectives. International Journal of Science Education, 10 (5), 541-552, 1988.
12. Hewson, P.: Epistemological commitments in the learning of Science: examples from dynamics. Eur.J. of Science Education, 7(2), 163-172, 1985.
13. Hewson, P. & Thorley, R.: The conditions of conceptual change in the classroom. International Journal of Science Education, 11 (5), 541-553, 1989.
14. Jimenez Aleixandre, M.P.: Los esquemas conceptuales sobre la Selección Natural: análisis y propuestas para un cambio conceptual. Doctoral Thesis, University Complutense, Madrid 1990.
15. Jimenez Aleixandre, M.P.: Thinking about theories or thinking with theories? A classroom study with Natural Selection. International Journal of Science Education, 14 (1), 51-61, 1992.
16. Jimenez, M.P. & Fernandez, J.: Selection or adjustment? Explanations of University Biology Students for Natural Selection Problems. In: Proc. Int. Sem. Misconc.& Educ. Strategies in Science & Math (Novak, ed.), Vol II, pp. 224-232. Cornell University, 1987.

17. Lawson, A. & Thompson, L.: Formal reasoning ability and misconceptions concerning genetics and natural selection. Journal of Research in Science Teaching, 25 (9), 733-746, 1988.
18. McClelland, J.: Alternative frameworks: interpretation of evidence. European Journal of Science Educ, 6(1),, 1-6, 1984.
19. Nersessian, N.: Conceptual change in science and in Science Education. Synthese 80, 163-183, 1989.
20. Novak, J.: Concept mapping: a useful tool for science instruction. Journal of Research in Science Teaching, 27 (10), 937- 949, 1990.
21. Pozo, J.I.: Aprendizaje de la ciencia y pensamiento causal. Visor, Madrid 1987.
22. West, L.,Fensham, P. & Garrard, J.: Describing the cognitive structures of learners following instruction in chemistry. In: Cognitive structure and conceptual change (West & Pines, eds.). Orlando: Academic Press 1985.

Food Relations of Living Organisms as a Basis for the Development of a Teaching Strategy Directed to Conceptual Change

Fred G. Brinkman

Universitaire lerarenopleiding Vrije Universiteit
De Boelelaan 1115, 1081 HV Amsterdam, The Netherlands

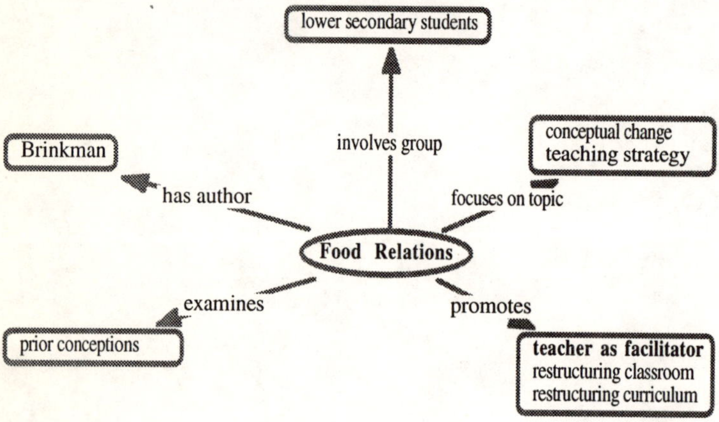

Abstract. One of the basic relations among biological organisms is the food relation. We investigated the biological knowledge displayed by students in lower secondary education when they performed a number of tasks during ecology lessons on food cycles. In these lessons a student work sheet was used, that we developed after the exploration via a concept-mapping test of students' conceptions before instruction about man, animals, plants and fungi and their mutual relations, to ensure an assimilative teaching strategy. Changes in students' conceptions during and after the lessons were registered via multiple choice tests (MC tests), and via interviews. The results show examples of students' concept development and of certain incorrect food relation ideas which seem resistant to teaching. The application of our results for the training of biology teachers is discussed.

Keywords. Preconceptions, food cycles, nutrients, dependence-relations in nature, plant-food, energy, plants, animals, man, fungi, bacteria, sunlight, students' drawings, students' scheme constructions

Introduction and Research Question

All biological entities require matter for their construction and energy for their activities. This is true not only for individual organisms but also for the communities that they form in nature. This flux of energy and matter through communities implies a dependency of organisms that is commonly represented in schemes of the cycling of nutrients through plants via animals, people and microbes and back again to the plants. We chose to formulate such a cycle in a lesson series after an inventory of ideas of students on the concept of 'interdependence'. Apart from its scientific significance, this concept has a central role in the ecology curriculum for lower secondary education in biology in the Netherlands.

A persistent preconception of biology students in lower and also higher education appears to be the idea that plants need organic food like carbohydrates and proteins to survive. This idea is in contrast with the unique photosynthetic capacity of green plants to use exclusively inorganic matter like CO_2 and H_2O as their sources of food.

In fact the preconception as such denies that animal and human life on earth depends completely on that unique process of 'organization' of matter that generates plant production and therefore the production of all life on earth.

This difference in perspective between scientists and children places the teacher in a position in which she or he needs, apart from considerable teacher skills, knowledge of pupils' present ideas and an understanding of their ways of processing information on this specific subject matter area [3].

The main question of this study is: Which differences in subject matter, knowledge and skills can be found when students perform domain specific tasks in relation with the concept of 'interdependence'?

During the research activities, lesson materials for biology lessons in lower secondary education were developed and their effects were registered. The development of these lesson materials was based on an inventory of student ideas on 'interdependence' before teaching, to ensure an assimilative teaching strategy. A multiple choice test was used to register possible changes in preconceptions of students (pre-post design). Some students were interviewed after the lessons to verify persistent ideas.

The nature of the study is explorative: written responses and protocols of interviews are analyzed and compared with the results of the different tests. The results of this study give examples of concept development that can be used by science teachers in their own classrooms. Moreover, the results can be used to confront student teachers with concept development in biology.

Context, Populations and Methods

Inventory of Ideas

Pre-instructional ideas were registered via a Hierarchical Concept mapping test (HC test) and analyzed as described earlier [2]. We added another aspect to the test in which the students had to find a connection between two different stimulus words.

The number of students is 53 (28 girls and 25 boys), aged 12–14, in two first classes of lower secondary education. The procedure included the following successive steps:

(1) Collecting associations (5 minutes): students were asked to write down individually words that they connect to their stimulus word: either man, animal, plant or fungus.
(2) Clustering of associations (5 minutes): students were asked to arrange their associations into what in their opinions are coherent groups.
(3) Name giving and structuring (5 minutes): students were asked to give a title to each of their coherent groups of associations and to describe the relation of each title with their stimulus word in a statement.
(4) Relations to the other stimulus words (5 minutes): students were asked to set relations to the other stimulus words, for example: "What has fungus to do with (a) man, (b) animals, and (c) plants?" The test instructions were given via a written handout.

The instructions and the time control were looked after by the teacher. The resulting ideas have been integrated into the lesson materials that were developed.

Multiple Choice Tests

This test was given before and after the lesson series. The test consisted of 21 multiple choice questions and the students have been asked to explain their choice of answer. The questions mentioned were partly new, and some questions were also used in earlier investigations [4]. A few questions were selected from Barker [1].

Lesson Series on 'Cycles in Nature'

The lesson series including the multiple choice pre- and post-tests took three lessons of 50 minutes. Forty-four students completed all three parts (21 girls and 23 boys). The students worked individually through self-instruction. The teacher didn't explain things to the students, except when indicated. The attention of the students was not always optimal, because of events outside the classroom. The Gulf War had just started.

Interviews

After a pilot interview, we structured the interview as follows:

(1) First, the students were asked to produce a 'cycle' with cards and arrows. On the cards were the names of plants, animals, fungus, man and non-living things like water, minerals, sun, energy, and soil.
(2) We asked questions about what they were doing and asked them to explain certain subjects they chose.
(3) We asked them to draw what happens to manure or dead leaves that are put on soil. First we gave them a drawn section of a piece of soil with this manure and leaves on top of the soil, and then a drawing of a plant with its roots in the soil with a flower and a lot of leaves above the soil.

The selection of students for the interview was based on some answers in the tests and some exercises in the lesson series, especially the parts that included the interpretations and the filling in of schemes about 'cycles in nature'. We selected four students doing extremely well in tests and lesson series and four students doing not so well. We made four couples of these eight students. Three couples were mixed and the other couple consisted of two girls.

Results and Major Claims

Components of the Nutrient Cycle

We took inventory, via a free word association test, of students' associations with four stimulus words: man, animals, plants and fungus (Table 1).

Man is described mostly by his anatomical, racial or other outer characteristics. The behavior in various occupations is the next component of the ideas of the students. Other headings are the course of life and relations between humans, especially family relations. The primary necessities of life, like food, form quite an important part of students' ideas about man.

Animals are living in a caged environment in the opinion of almost all students. Further, a number of species like domestic animals and parts of the animal body are mentioned. Food relations do not occur.

Ideas on plants are characterized by names of species or by anatomical features. The role of plants in food cycles and as a necessity of life is mentioned only by a small number of students.

Fungus is formulated as a decomposer of food, but its role in the woods is mainly 'living' without the indication of its decomposing role there. Other aspects mentioned are outer characteristics (color) and emotional reactions (dirty). As the Dutch word for fungus is synonymous with a horse specimen (a gray), this group of associations was also present in the bulk of student ideas on fungus.

Table 1. Categories of free-association responses to four stimulus words (examples of specific responses are given in parentheses): n = number of students responding, % = proportion giving each type of response

	Stimulus word: **Man** ($n = 14$)	%
1.	Outer characteristics:	
1.1	Parts of the human body (arm, leg)	50
1.2	Examples of different races	50
1.3	Other characteristics (sex)	43
2.	Occupations (work, walk)	64
3.	The primary necessities of life (food)	43
4.	The course of life (birth, death)	43
5.	Examples of family members (dad, sister)	36
	Stimulus word: **Animals** ($n = 13$)	%
1.1	Caged environment (zoo, house)	92
1.2	Free environment (woods, nature)	54
2.1	Animal species (in general)	69
2.2	Domestic animals (dog, cat, horse)	54
3.	Parts of the animal body (tail, leg)	46
	Stimulus word: **Plants** ($n = 13$)	%
1.	Plant type (tulip)	100
2.	Parts of plants (leaf, stem)	69
3.1	Necessities of life (water)	15
3.3	Necessities of life (sun)	23
3.4	Necessities of life (fertilizer)	23
	Necessities of life (CO_2)	15
	Stimulus word: **Fungus** ($n = 13$)	%
1.	Food becomes moldy (moldy bread)	62
2.	Fungus lives in the woods (mushrooms)	54
3.	A gray (horse)	54
4.	Fungus has different colors(green, blue)	46
5.	Fungus is dirty	46

Based on these results we selected the following assimilative topics to be included in the start of the lessons series under development:

(1) Occupations of humans in daily life situations.
(2) An animal example from the husbandry sector.
(3) A plant species, not a community.
(4) Food and out-of-door relations of fungi.

Interdependence Between the Components of the Nutrient Cycle

We asked the students to relate a stimulus word from their test to three other components of the nutrient cycle, always in propositional format. The results of this inventory are summarized in Table 2.

Table 2. Relations students formulate between the different components of the nutrient cycle. In this test the pupils were asked to write down the relation between their stimulus word and three other stimulus words. % = percentage of the relevant group of pupils making each connection.

	Relations between man and plants ($n = 27$)	%
1.	They are both living organisms	33
2.	Man depends on plant for oxygen and plant depends on man for other gas	22
3.	Man depends on plant for oxygen	15
4.	People eat plants	11
5.	People have indoor plants	11
	Relations between man and fungi ($n = 27$)	**%**
1.	People let food get moldy	33
2.	A human can get a fungus infection	26
3.	Fungus is dirty	11
4.	A gray is a horse (Dutch synonym)	11
	Relations between man and animals ($n = 27$)	**%**
1.	They are both living organisms	41
2.	Man and animals resemble a lot (Both mammals)	30
3.	People take care of animals	19
4.	People eat animals	11
	Relations between animals and plants ($n = 26$)	**%**
1.	Animals eat plants	54
2.	They are both living organisms	38
	Relations between animals and fungi ($n = 26$)	**%**
1.	They both live in free nature	35
2.	Animals eat fungi	15
3.	Dead animals rot, decay and mold	15
4.	A gray is a horse	15
5.	Fungus is a disease, animals can have that disease	15
	Relations between fungi and plants ($n = 26$)	**%**
1.	They are both plants (mushrooms = fungus)	38
2.	Dead plants mold and fungus clean up dead plants	15
3.	Getting fungus is a plant disease	15
4.	Fungus lives in or on a plant	12

As can be seen from these data, only a part of the relations between organisms are formulated in terms of dependence. Dependent relations are mentioned most frequently in the relations between man and plants and oxygen dominates there significantly. Only a minority mentioned the food relation between man and plants.

With the exception of 'animals eat plants' (54%), food relations form a minority of all relations mentioned as in: 'man eats animals' (11%), 'man eats plants' (11%), 'animals eat fungi' (15%). Between plants and fungi no food relation is formulated at all, although the cleaning up of dead plants as a function of fungus is mentioned (15%).

Aspects of these results were incorporated into the lesson materials in various ways such as:

(1) A specific domestic animal (cow) that eats a specific plant (grass) is chosen as an example to introduce 'cycles'.
(2) Human dependence on plants is illustrated. We stressed the production aspects as with wood products, but also food.
(3) Dependencies are introduced as food relations. References to dependence of oxygen have been omitted. We chose to omit oxygen initially in the lessons, because of the known difficulties students have in distinguishing the concept of respiration with the concept of photosynthesis. We planned to introduce the respiration concept later on.

Table 3. The teaching strategy: the structure of the lesson series

Event	*Focus*	*Comment*
pretest	cycles in nature plant metabolism	multiple choice questions with commentary
orientation	products of plants human needs for plant products human dependence on plants	
concept	structuring of associations of students on mapping 'cycles' focus on food-relations in nature	
information	cow dung, grass, soil microbes, molds, and their food relations	
application	open questions scheme drawing tasks	
posttest	cycles in nature plant metabolism	identical to pretest

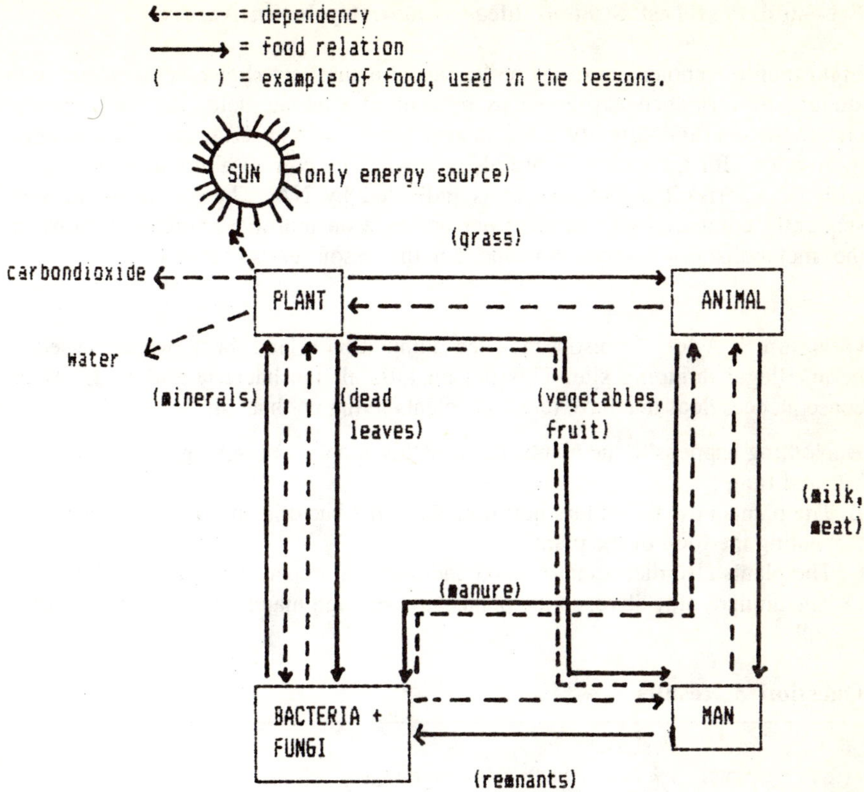

Figure 1. The teaching strategy: the content of the lesson series

The Content and the Structure of the Lesson Series

Based on the results of the inventories described and concerning other results [4] we developed a short lesson series for ecology lessons in the form of working sheets for students according to a teaching strategy we described earlier [3]. Table 3 shows the structure of the lesson series, Figure 1 shows the basic content of the lesson series.

We registered the participation of the students in the lessons by observation and via the working sheets, on which they had written down answers on text questions and tasks, without further instruction by the teacher. The effectiveness of this condensed form of instruction appeared to be 50–60% for all parts of the structure of the lesson series. Effectiveness here is used as the percentage of sufficiently answered questions in the eyes of the research team. The scheme drawing tasks had a lower score of 30%.

Pre- and Post-Test Student Ideas About 'Cycles in Nature'

Eight multiple choice questions on 'cycles in nature' are shown below along with student performance expressed as percent of students selecting each answer choice before and directly after instruction (n = 44). Examples of students' motivations for their choices are also shown. The correct alternative is flagged with an asterisk (*). No answer is indicated by NA. These results are subsequently compared with student performance on multiple choice questions on the 'metabolism' – a topic not included in the lesson series (Table 4).

Question 3. A lot of poison has been dumped on a land, because there used to be an illegal dumping site. This poison kills all the bacteria and fungi. What consequences does that have on all the plants living on that soil?

A. Nothing happens to the plants, because this is only a poison against bacteria and fungi.
B. The plants now have more nutrition, since the bacteria and fungus stopped eating the food of the plant.
C. The plants also die, because fungi and bacteria prepare food for the plant.
D. The animals stay alive, so the plants have enough manure to go on living after all.

Question 3 Results

Pretest (% of students)	Posttest (% of students)	Motivation
5	0	A. Because they need extra food.
9	2	B. They have more food and no more trouble with fungus.
84	91	*C. Bacteria make the food for plants. Poison is dangerous, it kills plants too.
2	7	D. I know that from biology.

Question 5. What happens to the manure and the remnants of dead plants and dead animals?

A. They'll be directly absorbed by the roots of the plant.
B. They'll disappear in the soil.
C. They'll be decomposed by bacteria and fungi.
D. They'll dry up to dust and then they're blown away.

Question 5 Results

Pretest (%)	Posttest (%)	Motivation
9	4	A. Manure is good for plants.
9	7	B. It rots slowly away.
71	80	*C. This is food for bacteria.
11	9	D. They dry out.

Question 10. Below is a sequence through which food is [nutrients are] passed. Number one is a green plant.

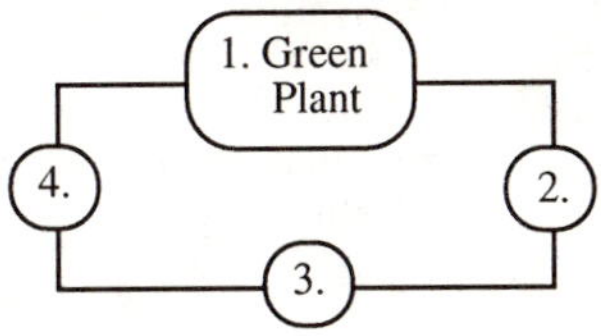

What is represented by the numbers 2, 3 and 4?

A. 2 cow, 3 manure, 4 fungus.
B. 2 manure, 3 fungus, 4 cow.
C. 2 fungus, 3 cow, 4 manure.
D. 2 cow, 3 fungus, 4 manure.

Question 10 Results

Pretest (%)	Posttest (%)	Motivation
64	64	*A. That is the cycle. The cow eats grass, makes manure, in which fungus lives.
9	11	B. The cow eats the latest.
9	14	C. The plant eats fungus and the fungus eats the cow and the cow produces manure.
18	11	D. The cow eats the plant and fungus makes the manure digestible (for the plant).

Question 11. In the cycle below the plant is removed. What are the consequences for the other parts of the scheme?

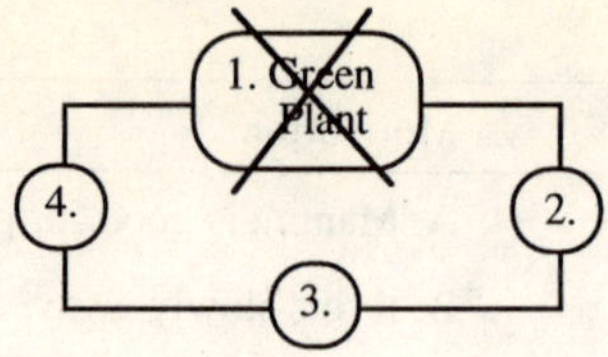

A. They'll adapt to another way of living.
B. At last they won't be able to live.
C. Number 2 dies, but number 3 and 4 won't notice a thing.
D. No consequences, the plants aren't important in this cycle.

Question 11 Results

Pretest (%)	Posttest (%)	Motivation
30	16	A. The animals make manure and plants live from that because adaptation is always possible.
46	71	*B. The plant takes care of everything. They all need the plant.
16	5	C. The cow stays alive and she makes manure.
2	9	D. The cow can easily live without plants.
7	0	NA

Question 8. Would plants through their roots be able to absorb all substances out of manure and remnants of dead animals, by themselves?

A. Yes, but only out of manure.
B. Yes, out of manure and out of remnants of animal bodies.
C. Yes, but the soil has to be humid.
D. No.

Question 8 Results

Pretest (%)	Posttest (%)	Motivation
16	9	A. That is already weak and can be soaked in the ground.
25	16	B. That is what they are living from.
21	16	C. They can not suck it up.
36	59	*D. That is what bacteria do. It has to go into the soil first.
2	0	NA

Question 18. What do plants take up out of the fertilized soil?

A. Water and manure.
B. Water and materials out of the decomposed manure.
C. Substances with a little bit of energy.
D. Water, manure and particles of soil.

Question 18 Results

Pretest (%)	Posttest (%)	Motivation
27	23	A. That is what a plant needs.
27	41	*B. That is what a plant needs.
21	18	C. That is their food.
25	18	D. That is what plants need. That is the smallest, the roots can take it in easily.

Question 1. The grass that a cow eats contains energy. Where does the energy in the grass comes from?

A. From the substances the grass takes in with its roots.
B. From the substances the grass gets from the bacteria.
C. From sunlight.
D. Because plants make their own energy.

Question 1 Results

Pretest (%)	Posttest (%)	Motivation
55	46	A. A plant gets its food from the soil.
9	39	B. Bacteria contain energy. Bacteria make minerals.
18	11	*C. A plant needs sunlight for growth.
18	5	D. Because it grows by itselves [sic].

Question 20. "Milk is good for you". This is a well-known saying in Holland. Milk contains a lot of energy-rich substances, proteins and fats which are important for the growth of your body. Milk is made by cows from their food, grass. Are there energy-rich substances in the grass which the cow transforms into substances in milk?

If not, how does the energy get into the substances in the milk?
If yes, how do the energy-rich substances get into the grass?

A. No, the cow takes in energy-poor substances of the grass and transforms them into energy-rich substances.
B. Yes, the cow takes in energy-rich substances from the grass and the grass had taken those substances from the soil.
C. Yes, the cow takes in energy-rich substances from the grass and the grass had made these substances out of energy-rich substances that have been taken in.
D. Yes, the cow takes in energy-rich substances from the grass and the grass had made these substances out of energy-poor substances that have been taken in.

Question 20 Results

Pretest (%)	Posttest (%)	Motivation
16	18	A. I heard that once. Because it is made by fungus.
46	55	B. The plant gets it food (the energy) of the soil [sic].
30	18	C. That's the way it is.
7	5	D. The plant gets the substances of people's manure.
2	5	NA

Table 4. A summary of student motivations for their response choices on the pretest and posttest about the subjects of cycles and metabolism. It illustrates in more detail, via the categorization of all motivations given, what already has been described above. (* = right alternative, NA = no answer).

Examples of motivations for student responses		Metabolism		Cycles	
		Pretest (%)	Posttest (%)	Pretest (%)	Posttest (%)
1.	No usable motivation.				
1.1	Guess.	28	16	86	35
1.2	I think; I have read;...etc.	28	27	78	99
1.3	Repeating the answer/question.	2	1	8	7
1.4	No motivation.	1	11	10	20
2.	How the plant gets its energy:				
2.1	Through the roots from the soil as:				
2.1.1	nutrition.	7	6		
2.1.2	manure.	1	2		
2.1.3	water.	10			
2.1.4	substances.		5	1	
2.1.5	minerals.		3		

Table 4. (continued)

Examples of motivations for student responses		Metabolism		Cycles	
		Pretest (%)	Posttest (%)	Pretest (%)	Posttest (%)
2.2	Not through the roots, but:				
2.2.1	through sunlight.	1	3		
2.2.2	other than sunlight.	2	2		
3.	Plants take in:				
3.1	Minerals.		2		
3.2	Water.			3	3
3.3	Water with a little bite of minerals.			1	1
3.4.1	Manure and dead remnants.		2	11	5
3.4.2	Manure but no meat.			1	
3.4.3	Substances of human shit.		1		
3.4.4	Roots do not absorb garbage.			1	2
3.5	Food and nutritious substances.	1	1	1	1
3.6	Things out of the humid soil.			3	7
3.7	Substances with energy.	2	-	1	1
3.8	Bacteria.			-	1
4.	Cycles				
4.1	Completed cycle (including decomposer).			7	23
4.1.1	No feedback from decomposer to producer.			9	2
4.2	Cycle without reducer.			1	1
4.3	Description of a food chain.			5	4
4.4	Non-logical cycle.			9	8
4.5	Plants depend on bacteria.			5	12
5.	The function of the decomposer in the food cycle:				
5.1	Only the decomposition of dead material.			9	16
5.2	Only the production of minerals.			-	4
5.3	Only the production of energy-rich substances for the plant.	5	8	1	1
5.4	Only the decomposition of dead materials and the production of minerals.			2	6
5.5	Bacteria eat:				
5.5.1	Plants.			3	
5.5.2	Dead nutrition.			4	
5.6	Bacteria and fungi are mostly situated on mold things.			3	2
5.7	Decomposers take care of the decay of dead material and produce manure.			1	1
6.	Poison kills everything and it is aselective			13	5

The following phenomena can be registered about student ideas on 'cycles':

(1) In the opinion of the students both before and after the lessons, elimination of bacteria and fungi in the soil has a deadly effect on plants. Sometimes a contradiction is found in their motivation; for example, the poison that killed the bacteria and fungi in question 3 is declared to be poisonous for plants, too.

(2) Manure and the remnants of dead organisms are decomposed by bacteria and fungi, is the idea of the students before and even more so after the lessons. After the lessons, about 16% of students, however, think that manure and organismal remnants disappear in the soil or by the wind (question 5).

(3) Some students discern two types of manure (see motivation, question 10D): the one that a cow makes, and another one, prepared by fungi, for the plants. 40% of the students keep – also after the lessons – the opinion that plants can absorb all substances from manure and dead organisms (question 8); the decomposition of manure is not, in their opinion, a necessary step (question 18).

(4) Elimination of plants has – before the lessons – for less than half of the students no dramatic effect on the life of the other participants in the nutrient matter cycle. After the lessons still about 30% of the students hold that opinion. The adaptability of life ranks as high as the alternative (question 11).

Although micro-organisms in the soil and their function in the cycles in nature are introduced in the lessons, their functioning remains relatively unclear, just as the role of manure in the cycle remains relatively unclear. Another concept that is not held by all students after the lessons is the essential role of the plant in the cycle of nature.

Pre- and Post-Test Student Ideas About Energy Metabolism in Plants

Some ideas on plant metabolism have come forward via some questions about energy, although the role of energy in cycles has not been stressed in the lesson series. It seems that the introduction of the role of soil bacteria in nutrient cycling has been accompanied by the idea that those microbes are responsible for the energy input into plants (question 1).

Sunlight as an energy source for plant life – although (briefly) described in the lesson text – plays almost no role in the opinion of the students. This is eliminated by failure to recognize the uptake of energy-poor substances by plants (question 20) by almost all students before and after the lessons. This result fits into existing preconceptions about plant feeding.

The lesson series is only partly effective, especially on the description of cycles as a whole. About the function of the components in the cycle with regard to the nutrient and energy flow, many a concept needs further teaching.

This is not very surprising because the time we took to introduce this subject in the classroom has been chosen to be deliberately short, to obtain information on questions or tasks that gave relatively low results. One task that gave such problems appeared to be the scheme drawing task.

Diagrams of Cycles in Nature

Cycles in nature are commonly represented schematically as a diagram like the one in Fig. 1. The cycle form stresses that the flux of chemical nutrients, the building blocks of biomass, are recycled by simply changing the form of the molecule of which they are part (e.g., nitrate-N —> protein-N —> nitrate-N).

Unlike the energy of solar radiation, which follows most of the time the same pathway as the nutrients in living communities, the nutrients are not in unalterable supply and the process of storing some in living biomass reduces the supply of nutrients remaining in the community. If plants and their consumers (animals, man) were not eventually decomposed, the supply of nutrients would become exhausted and life on earth would cease. In this respect, the decomposer system (bacteria, fungi) plays a role of overwhelming importance. It is this system that is least understood by students in secondary education. In the preceding paragraphs examples of the existence of this 'black hole' have been presented.

After the introduction on cycles in the lessons, we asked students to draw an example of a nutrient cycle of their own choice in their work sheets during the lessons. Some examples are shown in Fig. 2.

As far as this task was completed (the success rate was 30%) students draw three types of cycles:

(a) water cycle (Fig. 2a),

(b) a life cycle (Fig. 2c),

(c) an interrupted and incomplete nutrient cycle (Fig. 2c).

In the interview we held with some students, all examples of cycles made by them were of the incomplete nutrient cycle types (see Fig. 3).

Some general remarks:

(1) Arrows in the diagram are used inconsistently by the students (e.g., Fig. 3b).

(2) The decomposer system in the soil is present, but its role in the nutrient transition is unclear (e.g., Fig. 3a–c).

The nutrient transition system in the soil appears also to be unclear for students when they are asked during the interview to describe the fate of dead leaves and of dog dirt deposited on the soil surface.

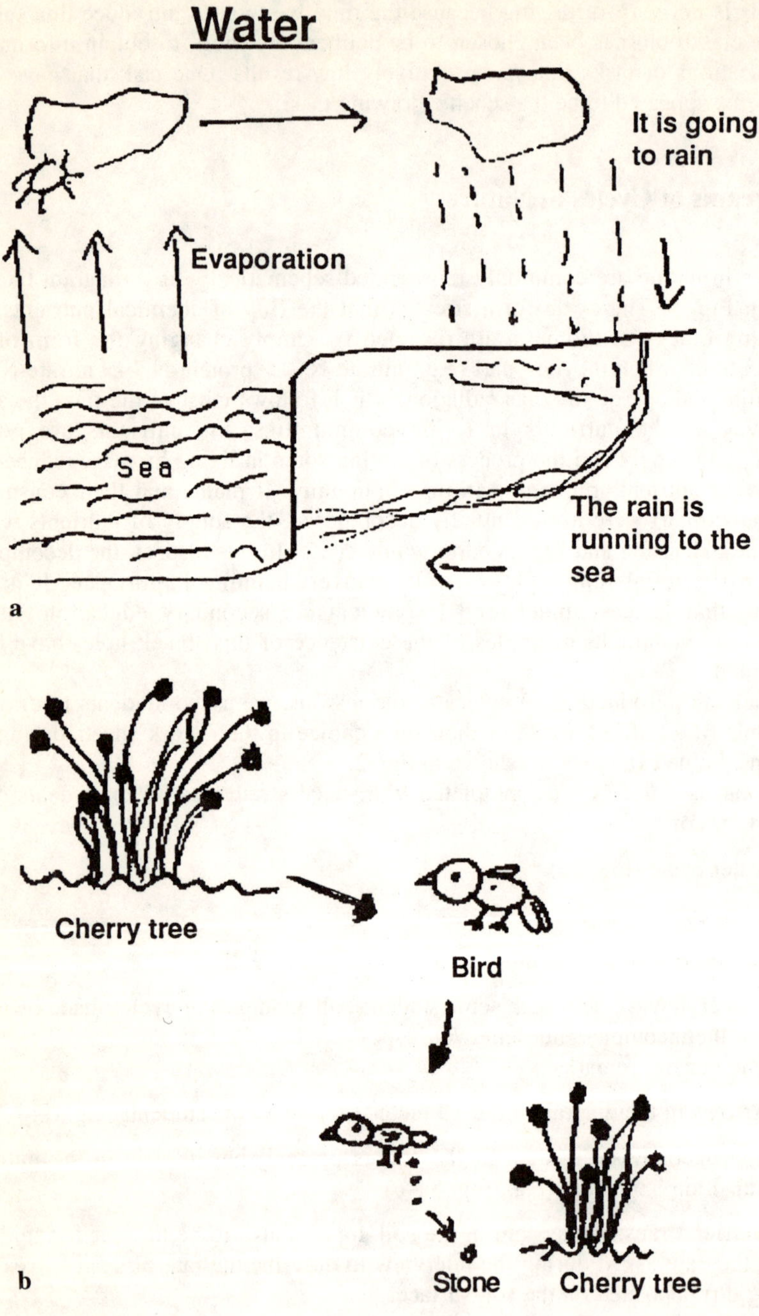

Figure 2. Nutrient cycles drawn by students in lower secondary education: (**a**) water cycle, (**b**) a life cycle, (**c**) an interrupted and incomplete life cycle

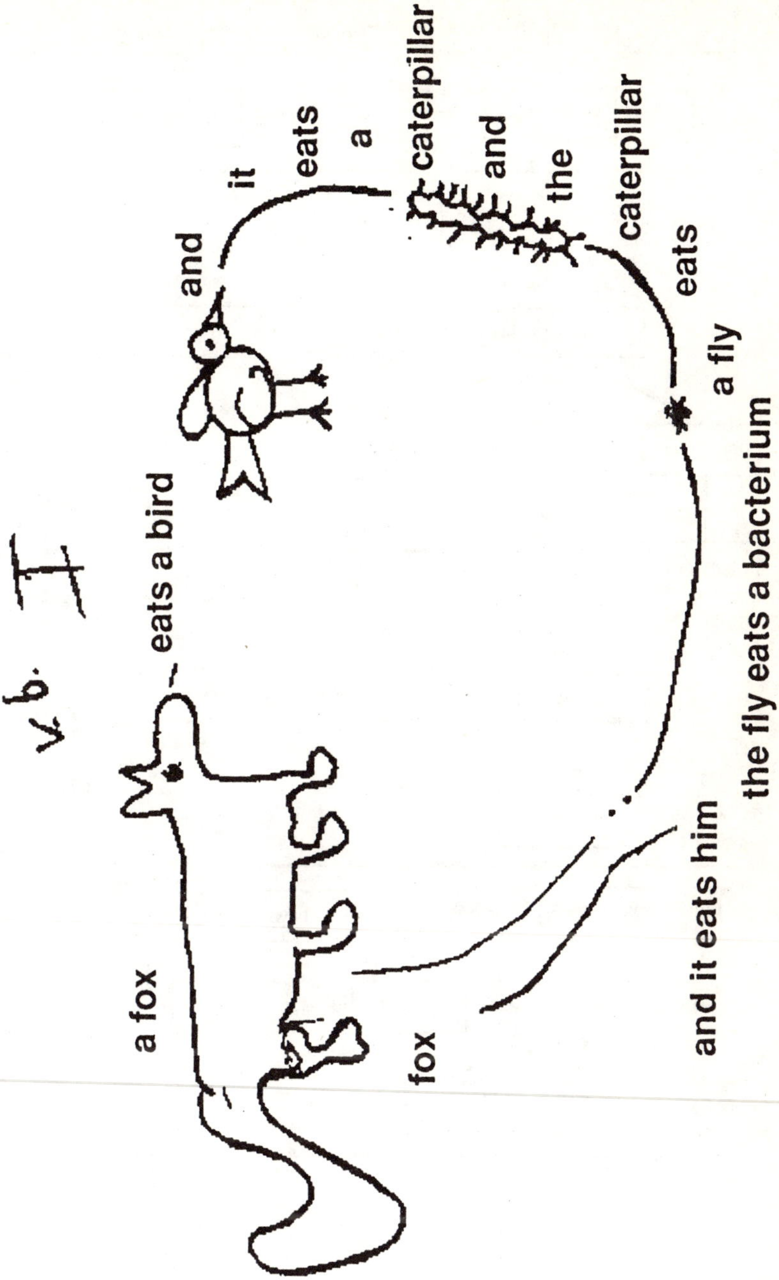

Figure 2c. *(see previous page)*

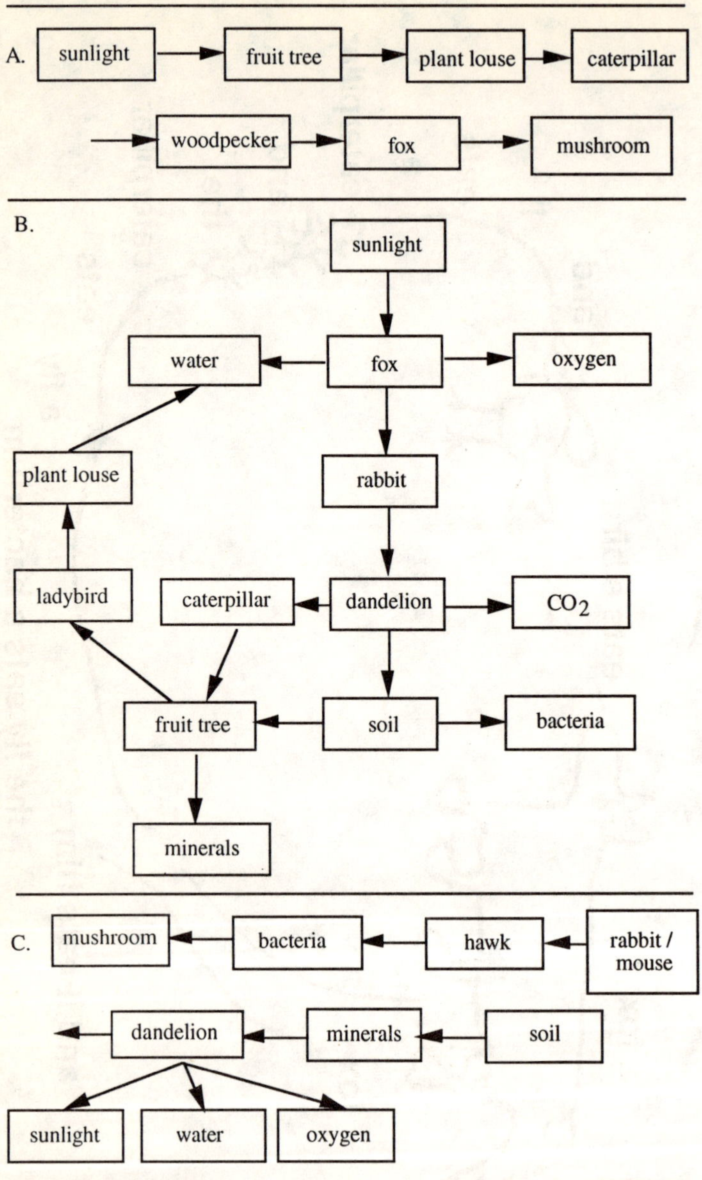

Figure 3a–c. Nutrient cycles, constructed by students with cards, on which a name of a component of a cycle has been written. Links = is needed by

Perspectives and Implications for Teaching

What has teacher training to do with all this knowledge, assembled in the classroom of pupils? There is a general remark to be made here. Although each individual has a unique set of past experiences, there is a remarkable consistency among the naive conceptions about specific science subjects that students in various cultures bring into the classroom. The task of the teacher is to unravel individual strands of ideas, label them and put them into a new fabric of understanding, almost without replacing earlier conceptions. This means for teachers:

- Don't deny the relevancy of students' ideas; make them visible and discuss them in relation with scientists' ideas.
- Be aware of the specific conceptions students hold, e.g., on topics like plant feeding.

Teachers acknowledge the problems students have with the subject matter of ecology. In a round table discussion with experienced biology teachers in Amsterdam (1991), the teaching of ecology came forward as a difficult task because students don't like the subject matter and judge it as insignificant. The reasons given for this were threefold:

- The structuring of the subject matter is not easy because of the great number of vaguely defined concepts.
- Applications of ecology concepts in real life situations are rarely given in textbooks and are not easy to find in the surroundings of schools.
- Conceptual problems of students are acknowledged around a series of subjects that have a relation with ecology: photosynthesis, respiration, matter and energy, evolution, adaptation, sustainable development, environmental pollution, level of organization (ecosystem, organism, cell), etc.

Lesson material that is written to cope with these problems can be used in training situations in pre-service or in-service courses to let (student) teachers become aware of the principles of knowledge as a restructuring process that can take place in the classroom.

Acknowledgment

The contributions of Jeannette van der Sluijs and Marco Lakeman to this study are greatly acknowledged.

References

1. Barker, M.: 'Where does the wood come from', University of Waikato, Hamilton, New Zealand, 1990. (A lesson series on plant photosynthesis).
2. Boschhuizen, R., Brinkman, F.G.: The biology teacher and research methods on pre-instructional ideas in education. (J.T. Voorbach, L.G.M. Prick, eds.) Teacher Education, 5, 65-75. 's Gravenhage: SVO/ATEE, 1989 (Chapter in a book).
3. Boschhuizen, R., Brinkman, F.G.: A proposal for a teaching strategy based on pre-instructional ideas of pupils. European Journal of Teacher Education, 14 (1), 45-57, 1991 (Chapter in a book)
4. Boschhuizen, R., Lakeman, M.H.: The influence of teaching the concept of decompose on the development of the concept of 'plant nutrition'. Teacher Education, 7, 17-30, 1991

Cognitive Strategies in Biological Thinking

Maria Arcà

Centro Acidi Nucleici del CNR, P.le Aldo Moro, 5, 00100 Roma, Italia

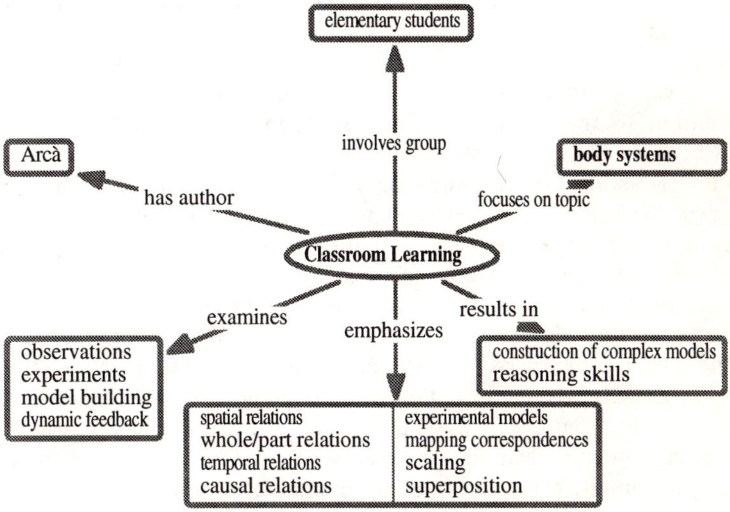

Abstract. Children's first attempts at explaining biological events point to original thinking strategies highlighting aspects of space, time and causality in the phenomena. Biology teaching in the school should work towards developing and integrating different kinds of children's thinking activities about biological facts, stimulating their growth and their dynamic re-organizations. The aim is to gradually bring the nativity of childish models up to the complexity of scientific ones. Discussions, observations and direct handling may produce a gradual replacement of children's simple ways of interpreting complex phenomena, developing more effective, varied and hierarchical strategies for understanding. A progressively more scientific way of looking can easily be elicited in children's interpretations, taking into account the net of interwoven causalities which supports biological complexity. On the basis of tape-recorded discussions and other children's products, I have been looking for the evolution of children's interpretations of biological phenomena, following the development of their cognitive strategies through long-lasting class activities.

Keywords. Phenomenological approach, cognitive strategies, knowledge dynamics, children's models

Objective: Teaching as a Cognitive Interaction

Many theoretical reflections have emerged from my research which aims to sensitize teachers to the various cognitive strategies always involved in children's construction of knowledge. In particular, I have been studying children's ways of accounting for biological phenomena. For several years, I followed children's cognitive and scientific evolution, through weekly programmed interventions in the classes, interacting directly with them and continuously checking for existing ways of thinking.

In diverse class activities, it is easy to notice similarities and diversities in children's ways of thinking, suggested and induced by observation. If teachers pay attention to these aspects, they can become adept at identifying the difficulties (epistemological obstacles) children must overcome to succeed (or fail) in understanding concrete situations. The examples I give here may suggest to teachers different ways to interpret what children say about facts; and they could get a better understanding of how different explanations emerge from the background of children's personal cosmologies. On the basis of this awareness, the educational work can be directed to promote an increasingly close fit between concrete children's experiences, the languages suitable to represent them, the common and personal ways of thinking, and the related processes of knowledge growth. My claim is that each learning process becomes effective only if it is able to activate dynamic feedback between the observations, their cognitive representation and their integration into shared knowledge (culture).

If we are committed to modify teaching strategies in order to give sense to children's biology education, we must take into account that each teaching-learning interaction requires coping with a multiplicity of problems, where at least three dimensions are systematically interwoven:

(1) **Cognition:** Which categories and strategies of thinking are active or must be strengthened in children's understanding of everyday biological phenomena?

(2) **Education:** Which concrete activities, experiences and experiments are able to support consistent interpretations of biological processes? How do teachers make the common biological models believable to children? Which kinds of interactions must be developed between children's and teachers' ways of looking at facts and intrinsic nature?

(3) **Curriculum:** Which new structures must be given to the biological subject matter to make them more consistent with observed reality, and with children's original thinking strategies? How to introduce children to a well linked net of correlated biological information?

On the basis of my research, I see that biological phenomena have the power to suggest biological models of understanding to children, teachers and researchers. Knowing can be seen as a human physiological process, involved in a flexible, dynamic reciprocal feedback structure. At the same time, models of biological explanations can easily be related to more general models of cognition. In this sense, each suggestion of patterns in knowledge construction has to satisfy both epistemological and biological requirements.

The perspectives offered by this research carry strong implications for teachers: it is worth while for the teachers' (but not easy) to become aware of the cognitive ways in which children construct their scientific knowledge. If teachers can agree that any knowing corresponds to a dynamic process, and not to a thing possessed to some degree, they can easily guide children to transfer the cognitive strategies mastered in a particular context to different ones. Teachers committed to this process can play an effective role in the complex interaction that always involves, at the cognitive level, adults and children, experience and knowledge, languages and representations.

Interactive Methodologies: A Phenomenological Approach

In real class situations, across the five years of the elementary school, I have suggested several themes for discussion, starting from observations and manipulations of concrete biological objects. Interpretations of long-lasting phenomena, such as the development and the growth of living beings, were coupled with others related to short-term activities driven by the learning context, such as the observation of biological phenomena and objects, performing specific experiments, and the suggestion of new hypotheses by the children. By means of different kinds and levels of questions I stimulated children's interpretations of complex biological phenomena, looking for the meaning, range and limits of their original ideas.

This work was supported by the analysis of tape-recorded discussions and other children's products. Possible restructuring of subject matter areas was further explored through the intervention itself, and systematically reorganized a posteriori. The aim was to achieve efficient coupling between children's cognitive evolution and science-driven reinterpretation of natural phenomena.

Introduction: Knowing by Models

In society, we see that children's and adults' knowledge is shaped by observation of nature facts and the many interpretations given through common sense and our scientific culture. For children and society, modeling and inferring are unavoidable and inseparable aspects of cognitive and scientific construction [1]: they are part of an endless game of reinforcing, modifying and adjusting mental models to reality. This systematic approach, proceeding by trial and error, fitting, invention and successive approximations, defines a general way to knowledge construction, which starts from what is known, and is continuously engaged in readapting old knowledge to the new. So, even for children, any new understanding starts from what they know: not only from what they directly perceive of themselves and of the surrounding reality, but also from what they indirectly know through their surrounding culture. We all live in a natural as well as a cultural medium, and ever since birth we experience both as mutually compatible: inside and outside of school, children use both these sources of cognitive experience as a basis for the construction of new personal knowledge [11].

In children's cognitive constructions, as in the social development of scientific thinking, all knowledge about the world refers to different sets of models of the world. Within this process, schematized aspects of reality and of thought processes are continually selected and reciprocally fitted [2, 3]. Every act of knowing is therefore a partial one, in extent, structure and meaning, according to the quality and the status of the fit that it represents. Following different aims, different fits can be selected, where aspects of direct experience, representative-symbolic language, organized knowledge, and thinking strategies are intertwined [6]. All knowing therefore refers to other forms of knowing, i.e., to other correlated networks of partial fits. At the same time, all knowing is temporary, undergoing continual evolution; and each cognitive strategy corresponds to a particular way of looking.

As teachers, we can guide children to make their original and naive models fit better to changing realities, and to establish reciprocal links [9]. No one possesses a priori cognitive frame (models of models) [12] or is consistently able to organize notions and observations on specific aspects of reality. We can suggest such scientific models to children when they are ready to assimilate them as a consequence of having acquired the general frames to support them.

This account of the construction of complex structures from simpler ones quite easily leads to paradox [4]; however, one of the cognitive strategies which appears to be able to "reach the more starting from the less" is the analogy mode. It is a partial formal similarity between two or more representations: one of them, the one related to better known aspects of reality, supports the other in organizing, then representing what is not yet clearly grasped. Thus, reasoning by analogy gives children the facility to compare models, to compare schematized real situations, and, finally, to construct new models, which extend and reinterpret the initial ones.

Strategies of Analogical Thinking

In a normal class, children (7 years) always converse among themselves and with their teacher. Whenever they take up a new activity they compare their personal opinions on a specific topic: this is the starting point for the observations and experiences which will follow. In this study, the topic was: "how does the eye work, and how is it related to visual perception?" Each child attempts to give an explanation of what happens while looking at a particular object. How can we see anything? Which mental processes, which physiological activities, which internal structures, which physical features are necessarily interlinked in such an astonishing phenomenon? Things are there but, if one sees them, they get into one's eyes, or into one's brain...not the object itself, but a picture of the object...an image of the object, like a photograph of the object. Of course, the eye is like a camera...But we think with our brain, and it is our brain which gives 'orders' to the eye: the brain orders the eye to see ... the eye takes the photo and the brain acts as a developing-machine; it sees the photo ... and we see the object. It is clear.

In these discussions, we see how analogical models act when they try to explain unknown processes: the coherent development of the first-order analogy is used by the children to account for the whole process. The modeling procedures always attempt to connect children's direct experience with the re-elaboration of cultural tracks, transmitted from adults' to children's knowledge through linguistic channels. Now the question for the teacher is: what degree of reality do children give to their own models? In other words, what is the meaning – for instance, in this situation – of the likeness children evoke between an eye and a camera, as it is clearly impossible to imply a physical reality for a photo-taking and developing machine into the brain? How then can an adult, a teacher, go into the significance ascribed to the multiplicity of analogies which children usually produce and handle, in their effort to explain some otherwise unaccountable aspect of reality? Conversely, what is the meaning children attribute to the great many of analogies adults, teachers, textbooks and media continually use in 'talking about' and 'talking over' facts and phenomena? And, if children actually take as a reality what is an adult-suggested analogy, how to discover that? How to convey at an analogical, abstract, level the often complex picture which children believe to be a true matter-of-fact description?

Strategies of Representation

In the multiplicity of children's cognitive reconstructions of events, we can recognize many different ways of looking at complex situations, i.e., different strategies, but each childish model of reality is generally guided by just one fixed strategy which partially interprets facts and phenomena according to a unique, very specific, point of view.

So we can recognize strategies which are able to look at spatial relations between systems, or between parts of a whole, the emphasized features are therefore mainly shapes, positions and localization's. A 'correct' model of this kind accounts for how and where objects, or their parts, are settled within a given framework: sometimes it includes a topological organization of their relationships. Most naive models of the human body belong to this spatial type and I find them very similar in 4- and 12-year-old children. Nobody can be amazed that such representations are often *not* substantially modified with age. Often pupils are required only to memorize names and positions of different parts, without any suggestion of new dimensions of knowledge powerful enough to arouse and sustain more complex modeling. In many textbooks the only connotation of organs and apparatuses is by their shape and position, within a body treated as a container of organs [5].

In their representations, children also adopt temporal strategies as when they use space (on their sheet of paper, for instance) to show a succession of biological events such as growth, transformation, development. In these cases, the cognitive problem is twofold, on one side, how to transform the understanding of a diachronic sequence into a schematic synchronic form by means of a representation; on the other hand, how to convey, within a, linear time-scale, the several aspects of any functionality which needs to be interpreted as a multiplicity of concurrent functions, each of them proceeding at its own rate.

Problems in giving meaning to representations also arise in the handling of experimental models. It is clear to a teacher, as well as to a scientist, that any experiment corresponds to a schematic modeling of events, and aims to reconstruct only a few, salient aspects of a more complex pattern. Many questions asked while events are happening call for experimental answers. Any experiment is a reality-like event in a specific setting, while the results and the interpretations of the data always involve a priori abstract models of the real phenomenon as a whole. Guided by analogical thinking strategies, the results of experimental reconstruction can take their place in natural contexts. This is allowed by a kind of superimposition, over the whole, of artificial and schematic reconstruction which select only partial aspects of reality. Too often, for children, an experimental result is only the outcome of a procedure, and they do not recognize in the experimental operations the schematized pattern of a real event.

At school, since I aimed to teach children to perceive simple relationships between the morphological and physiological features of different organisms, I used to involve them in the task of anatomical recognition by observing parts of animals. At the same time, I suggested to them mechanical models of the physiological functions. Analogical thinking strategies in fact allow children to see how mechanical objects can mime organs or body structures responsible for single operations. So, a one-way hydraulic valve can be like an aortic valve: and the blood stream through the aortic valve can be modeled by the air stream blown by children through the hydraulic valve. Similarly, to explain the complex processes of assimilation, the passage into the blood of the 'good' parts of ingested substances can be modeled by analyzing the properties of various mechanical filters, which can allow the passage of more or less tiny particles of food ground by children. But in all cases children's problems are about the meaning of the experimental model. In the modeled situation, in fact, they need to know explicitly which part, object, function, process of reality is modeled by which part, object, function, process in the experiment. Working in the laboratory they often lack any clear rule governing the correspondence between reality and the suggested or imposed experiments. Local experiments have then to be joined together by abstract links; and many different strategies are needed to imagine a functional model of a living system.

Also for adults, strategies of scaling up and down, in time and in space, are all invoked to picture and to explain the different roles of macroscopic, cellular, and sub-structural features, and their complex interactions. Furthermore, any realistic knowledge of living systems has to be grounded in the cognitive reorganization of local processes at various levels of dynamic functional frameworks. Only a complex model of parallel processes of respiration, absorption and digestion, diffusion, circulation and excretion, at both whole organism and cellular level can account for the general features of life. There is no problem with this kind of modeling as an aid to understanding dynamic processes, changes and transformations. But what happens if, for instance, lower-middle school pupils are explicitly asked to express their views of the process of human digestion? Physiological activities occur according to necessary patterns; and any scientific modeling, supported by causal interpretation schemes, has to be consistent with events. Even pupils' own modeling cannot escape casual connections within necessary, even if unknown, processes. So, driven by what must happen, it is almost impossible to invent a fictional-model substantially different from the accepted one.

In class situations that I have studied, pupils tried to invent odd machines able to carry out the necessary complex patterns of a digestive activity, or else they attributed invented names to working parts and structures essential to fulfill necessary roles within their schemes. As a result, even the most personal models often reproduce, in their deep structures, well known conventional scientific features: but most often the imagined working machines are represented through static shapes and networks of shapes.

Strategies of Causal Thinking

Dynamic models of biological phenomena involve cognitive strategies of causality which lead to the identification of causes, changes of causes, modulation of causes, interaction among causes and so on. But the need for causes is embedded, in many ways, in children's explanations of the complexity of vital processes, and in their relationships. On these bases, in my research, I have been looking for the evolution of causal cognitive strategies. In different learning situations I have stimulated children's interpretations of biological phenomena; looking for meaning, range and limits of their causal thinking. Despite the evolution of scientific thought, determinism and finalism are both still rooted in common thinking and in the language used to speak about simple but stimulating biological events [7, 9] children's ideas are deeply shaped by the surrounding culture. Furthermore, simple observations of living beings, starting from seeds and eggs, do not allow children to understand the essential features of development and change, nor to imagine future shapes and stages. The variety of transformation processes, from the sprouting of seeds to the metamorphosis of amphibians and insects, cannot be simply explained by direct cause-effect links. Children often appeal to a global causality to account for the interwoven dynamics of growth. Moreover, in children's thinking, time often appears as the main cause of the happening of things. Growth, development, transformation ... thus occur only because time is passing.

In modern biology, [8], causality chains do not fit in with the complexity of biological phenomenology. Self-regulation and negative feedback allow living organisms to retain their characteristic features, in spite of environmental changes and ontogenetic dynamics. The biological understanding of life always implies different levels of causal links, including genetic and environmental correlations.

Faced with the complexity of biological events, children often attempt to explain the physiology and ethology of animals and plants in terms of an inclusive, genetic cause, invoking the very nature of that form of life. In this sense, children's interpretations are perhaps similar to ancient vitalistic approaches, in which life is referred to as the very cause of any particular vital process; but, perhaps they are not too far from modern systemic approaches, which refer to an underlying complexity to account for the phenomenology of life.

However, in the construction of appropriate dynamic models, causal strategies systematically attempt to correlate, explain, and predict biological events. But, it is easy to see that each kind of causal explanation produced by the children involves in turn a complex cosmology, an image of the whole living system. Often within its own environment, it is the background for what has to be specifically explained. In these pictures, the living systems live and work because of a kind of spontaneous vitalistic causality, which is a reliable and handy explanatory model for children. Discussions and observations of facts may produce a gradual replacement of such global, intrinsic causality with precise, local causal relationships. A more scientific way of looking can then be elicited in children's interpretations, taking into account the net of interwoven causalities which supports the biological complexity. Yet, in the processes of learning, some children's cosmologies and habits of correlation conflict with guided observations or teaching contents; which is often reflected in discrepancies in their causal thinking.

Scientific Models

In my experience, children's ways of looking at biological reality are often grounded on a single interpretative strategy at a time. Thus they use only spatial, or only temporal, or only microscopic ways of interpreting the complexity of living organisms. It is difficult for a child to combine different strategies at the same time, or to shape a model founded on more than one dimension.

Only reciprocal connection and superposition, and explicit thinking strategies can produce a cognitive reconstruction of reality, able to take into account its various features; ability to interact casually with, and to modify, any specific event, for particular purposes. Adults' models, cultural models, and in particular scientific models, are multidimensional reconstructions of aspects of reality. Only against an ever present background of different dimensions can a single one emerge from time to time, dominating the others, demanding priority for its specific way of looking at knowledge and events. In other words, only a range of contemporary cognitive strategies is adequate to model different situations, and relationships between situations. This availability gives humanity cognitive and factual power over reality, and allows the building of causal nets between knowledge and events. The richest strategy is modeling. The deepest could be the cognitive mastering of situations, capable of analyzing complex patterns of reality according to different criteria simultaneously. Reasoning in terms of causal complexity leads children to invoke more complex feedback processes and feedback controls to explain the specificity of life processes. Moreover, the generalization of feedback models, for instance at the cognitive level, offers teachers and researchers a powerful way to interpret the complex processes of knowledge construction, guided by looking for the best fit between scientific interpretations, intuitive cognitive strategies and reality.

References

1. Arcà, M.: Strategies for categorizing change in scientific research and in children's thought. Human Development, 27, 335- 341, 1984.
2. Arcà, M., Guidoni, P.: Scientific knowledge, cognitive development and basic science education. Proceedings of the USA–Italy joint seminar on elementary school science education. 1983, Frascati, Roma, pp. 29-41. CEDE Roma 1984.
3. Arcà, M., Guidoni, P., Mazzoli, P.: Structures of understanding at the root of science education: Part I: Experience, language and knowledge. Eur. Jour. Sci. Ed., 5, 367-375, 1983; Part II: Meaning for formalization. Eur. Jour. Sci. Ed., 6, 335-341, 1984.
4. Bereiter, C.: Toward a solution of the learning paradox. Summer, 55, 2, 201-226, 1985.
5. Caravita, S.: [Chapter in this Book]
6. Guidoni, P.: On natural thinking. Eur. Jour. Sci. Ed. 7, 2, 133-141, 1985.
7. Jimenez Aleixandre, M.P.: Thinking about theories or thinking with theories? A classroom study with natural selection. Int. J. Sci. Educ., 14, 1, 51-61, 1992.
8. Mayr, E.: Toward a new philosophy of biology. Cambridge, Massachusetts: The Belknap Press of Harvard University Press, 1988.
9. Strike, K.A., Posner, G.J.: A revisionist theory of conceptual change. In: Philosophy of science, cognitive science and educational theory and practice (R. Duschl, R. Hamilton, eds.), SUNY Press, Albany, NY. (in press).
10. Tamir, P., Zoher, A.: Anthropomorphism and teleology in reasoning about biological phenomena. Science Education, 75, 57-68, 1991.
11. Vygotski, L.: Thought and Language (E.Hanfann, G.Vakar, eds.). Chicago: The MIT Press 1962.
12. Wittgenstein, L.: Philosophische Bemerkungen. Oxford: Basil Blackwell 1964.

Organizing the Concept of Organism at the Elementary School Level: A Case Study

Silvia Caravita

Institute of Psychology, National Research Council, Rome, Italy

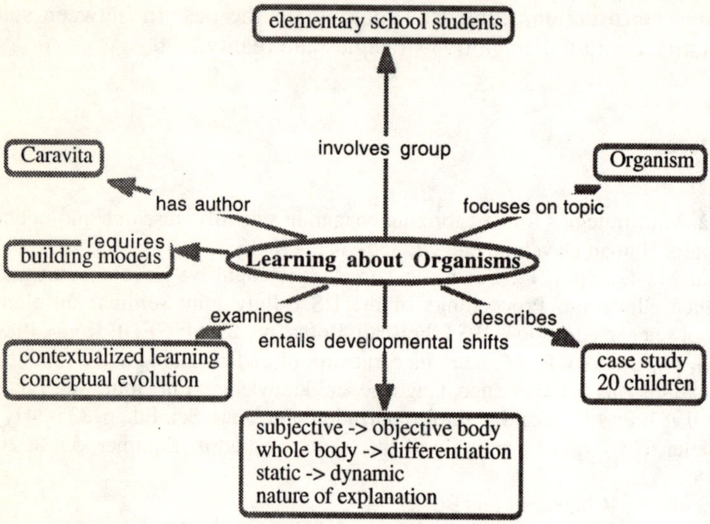

Abstract. Children's conceptions of the human body grow from a self-centered to an objective view, from a holistic to an analytical approach, which leads to a more differentiated conception, but which ultimately is no longer able to account for the complexity of the organism. The 'one part-one function' view is very likely to result from structure-oriented textbook descriptions of independent body components which remain largely unchanged even in adults. The construction of a systemic view is a long-term process which can begin in the elementary school. Our observations bear on a case study based on a 4th grade class, and was aimed at documenting the processes activated by various contexts where different symbolic media were used by the children to build models and make their ideas explicit. Each context triggered different reasoning strategies for explaining the relationships among the body's systems, and supported a range of negotiations among the children who had to assign meaning to the analogical constructions that they were producing. On the basis of these observations, the evolution of conceptualization in this area has been re-constructed. Rather than a dramatic restructuring of each concept, the data seems to demonstrate a diffuse change in important aspects of the mental ecology in which the concepts are embedded.

Keywords. Biology thinking, organism, conceptual evolution, contextualized learning

Introduction

Our basic assumption is that the biological world is a domain of knowledge which has a very specific status in our making sense about the world: we play a double role of 'the observer' and of 'the observed'. Cognitive as well as perceptual, interactional, emotional implications derive from this phenomenon and have to be taken into account when learning processes in this domain are explored. The introductory chapter in this book discusses this issue and no further arguments are needed here.

The numerous studies by Carey [6, 7, 8], Gelman R. [13], Gelman R. et al. [12], Gelman S. A. [14, 15] and Keil [18, 19] have investigated in depth conceptual changes occurring, along with cognitive development in the ability to categorize living 'objects'. Not many studies, though, have been concerned with the long-term organization and re-organization of children's interpretations of the facts and events which are addressed in school by the Life Science curriculum. On the other hand, students' naive conceptions about the physical world and their modification have been extensively investigated by educational and artificial intelligence scientists. In my opinion, such studies have greatly influenced the leading views of the process of conceptual change [22, 23] and they have been driven by epistemological theories on the rational construction of scientific knowledge again mainly in the physical sciences. Conceptions are modeled as mental structures whose semantic and syntactic components have been the main focus of attention. Children's ways of reasoning out explanations about a complex and unpredictable reality like biological phenomena can contribute to framing the process of knowledge growth as much as a complex, dynamic and 'biological' system, less mechanical than the engineered instructional strategies recommended by some science educators.

Tools for Biological Thinking

Rather than exploring children's knowledge in limited fields, more interesting is understanding their ways of viewing biological phenomena and how their reasoning is influenced by the features of the living objects and events and by the contexts in which they are interpreted. The conditions under which children use and acquire those very general thinking strategies that biologists value as being relevant to the comprehension of biological phenomena are our main focus of attention. Some of the more important tools of the biologists' thinking repertoire includes: a systemic view, dynamic rather than static interpretation, looking for patterns in variability, switching among levels of organization, shifting from the individual to the environment and vice versa, and selecting either continuous or discrete perspectives as appropriate.

We are also well aware that beliefs and emotions colour our interactions with living organisms, with one's own body and with other people around us, and with animals, and that their roots are in our culture. There are sources of resistance, therefore, which are not only phenomenological but are prejudged. Knowledge growth cannot be expected to occur according to the logical structure of the content only, since it depends significantly on the psychological meaning which a person assigns to it.

This report deals with children's interpretations of the life of their bodies. The key questions of the investigation presented here are: to what extent in the elementary school can we promote habits of thinking in terms of relationships among structures and functions? How can different social contexts and media for the representation of knowledge elicit different strategies for organizing the ideas? What evidence do we have of the processes which produce change?

The Study

Children's Conceptualization of the Human Body

In the Italian national elementary school syllabus the human body is a topic introduced in the 4th and 5th grade. Textbooks provide simple anatomical descriptions. The bodily functions are broken down into sequences of simple operations and each step is labeled with the scientific term. The graphic representations reinforce an idea of the body as a container of separate stereotyped pieces, each one having its well-defined shape, color and location. Where the blood flows in veins it is colored in blue and in arteries it is colored in red, but the continuities between them are not made explicit . No wonder then, that children's pre-existing conceptions are strengthened by instruction: the body continues to be conceived as a tunnel where food and air travel along, where the 'good things' are kept and the 'bad things' are eliminated.

The results of our previous studies [3, 4] comparing children from three age groups, from elementary to middle school, are in agreement with the main findings reported in the literature [1, 2, 9, 11, 16, 24]. Between 7 and 9–10 years of age two important shifts occur: the first is from an egocentric view of the body conceived as 'myself', whose functions are seen in terms of "my actions" or behavior, to a more objective view of my body as the human body. Presumably, cognitive maturity together with the experiential basis provided by the observation of other human beings and the animals which can be most easily likened to us, are responsible for the acquisition of this ability. The object-body can become accessible when the situation requires this kind of perspective, but the old perspective, the me-body, is not erased: even biologists can egocentrically approach living beings in their effort to interpret them, as the autobiographical notes by Barbara McClintock suggest [10, p. 117]. When she was making strenuous efforts to make sense of the events that she had repeatedly observed, she reports having got to the point of "really working with them [chromosomes], I wasn't outside, I was down there with them and everything got big. I even was able to see the internal parts of the chromosomes; actually everything was there. It surprised me because I actually felt as if I were right down there and these were my friends."

The second shift has rather to do with the inputs of school instruction, which mainly act at the linguistic level and provides a richer biological vocabulary. Children become less inclined to adopt a holistic conception of the body and show a growing awareness of its articulation into various inner organs. Also, each organ is no longer treated as the cause of a global function. Unfortunately, the acquisition of an analytical approach; and, therefore of a more differentiated conception does not result in an improved understanding in terms of the concept of organism at the end of the educational itinerary. The compartmentalized approach needs to be developed into a systemic view, and this does not seem to be the case for the majority of the students, even at the higher school levels [16].

There is another important aspect which is relevant to conceptualization in this area: the cognitive attitude towards the living acts as an epistemological belief and dictates what counts as a problem, what as an anomalous fact, what requires an explanation, what is acceptable as an explanation. In everyday thinking the living appear as self-explanatory: causal relationships are not sought to make sense of what happens, such as transformations, growth, movement, variability, etc. Children's ability to relate structure to function increased over the age span investigated. This type of causal reasoning seemed to be applied more spontaneously when the functioning of the artifacts is questioned. Inferring function from structure was found to be an easier mental process than the reverse. Familiar technology seems to offer the basis for making analogies, particularly when the knowledge available about the variety of animal structures is limited [4]. Gelman [13] provides data showing that children as young as age 4 realize that natural causes exist independently of human action or intention. Also, children apply different causal mechanisms to each natural event observed when prompted to do so. Children appreciate the significance of non-obvious properties for surface appearances and behaviors, and they realize that causal mechanisms may be inferred rather than directly observed. But, in spite of this cognitive disposition, they seem to imply that it is the constitutive nature of animate objects that enables them to move on their own. Psychological essentialism has been postulated as a general heuristic available to people whenever they lack the pertinent knowledge to make inferences about particular instances [21].

Are we so sure that the fragmentary, inconsistent image of the functioning of the body which emerges in the interviews, the drawings, and tests can be identified with the knowledge that children have? To what extent are the contexts of assessment responsible for the emerging features of that knowledge? There is no doubt that the culture in which children grow up is one source of their ideas and that it sets the historically determined frame inside which the whole range of possible actions, perceptions, verbalizations, interactions, and emotions dealing with their body is constrained. But there is also no doubt that from birth each child experiences the body as a source of information about reality. Each child 'lives' in the interaction with reality, learning about their own body as a consequence of the ways it reacts, it has to adapt, it resists, it hurts, and it changes. The child gathers and processes a great number of experiences which build up a rich, mainly tacit and continuously changing, knowledge that is fundamental to the various contexts of life. From these experiences the child absorbs many emotional overtones as well. When coping with the body at the formal level demanded at the school, children necessarily make up verbalized images from information that is beyond experience or from perceptions that are beyond words: the perceived digestion is something very different from the physiological process which goes by the same name. Verbal or graphic symbols are inadequate to express the simultaneous and multi-temporal complexity of the organism.

Therefore, the important role of educational intervention is to make the apparently simple phenomenology of life worth being questioned and investigated, and to provide students with opportunities for tracing back their intuitions and sensations. These may substantiate the formal knowledge whenever the evidence cannot be afforded by instruction, and conversely may help to challenge the students each time they are made aware of new information. This back and forth movement between the two resources of knowledge available, private and tacit on one side, public and explicit on the other, is the essence of the continuous process of change which should take place in school; it has to be instantiated within contexts where goals, media tools, and social interactions specifically shape student views and their expression. Figuring out and setting appropriate situations for the students is the hard but unavoidable task facing educators.

A Case Study

One class of 20 children in a school in suburban Rome was the target sample followed during the 4th and the 5th grades, to study long-term teaching/learning sequences. The data collected during the first year have been analyzed so far.

The life sciences curricular activities (Table 1) were discussed with the teacher. The tasks given to the pupils during their visits to our laboratory were also cooperatively designed. Tables 2, 3 and 4 display a detailed description of three of the tasks, to exemplify some of the contexts created for each activity. All these tasks demanded the elaboration of representations through the manipulation of different symbolic media and at different levels of abstraction. The following sources of data made up our protocols:

- individual written work-sheets filled in before or after the execution of tasks performed while working in a group;
- verbatim transcripts of the recorded verbal interactions occurring within the groups while performing the tasks;
- transcripts of the audio-taped presentations done by the children at the end of the group-work to illustrate the products to their classmates and of the class discussions accompanying the execution of the tasks;
- field-notes of the performances and contributions to the discussions of a sub-sample of eight children, which included boys and girls of various scholastic abilities, were summarized as individual conceptualization histories.

Table 1. Main topics and activities in the 4th grade curricular plan. Bold characters highlight the tasks described below in detail.

*** The structure of an animal body**
The rabbit's leg and its comparison with the chicken's leg (individual written description and drawing)
How does the rabbit move? (discussion)
Dissect the rabbit's leg
Collectively report on the observations carried out
Write individual report
How does the rabbit move? (discussion while observing a rabbit move)
Build a model of the rabbit's leg
*** Explaining functions: locomotion**
The rabbit's leg and its comparison with the chicken's leg (individual written description and drawing)
How does the rabbit move? (discussion)
Dissect the rabbit's leg
Collectively report on the observations carried out
Write individual report
How does the rabbit move? (discussion while observing a rabbit move)
Build a model of the rabbit's leg
*** The relationships among parts and functions in the body**
Assemble a concept map on a poster to explain what the organism is and what it does
Confront and interpret the posters
Revisit one of the posters one month later
*** Invertebrate structures**
Observe tiny animals under the microscope
Fill worksheets about invertebrates (individual task)
Observe Daphniae and their heart beat rate
Observe earthworms and their photo-sensitivity
*** The development of organisms**
Imagine and represent the development of a chicken embryo
Observe incubated eggs with the stereo-microscope
Interpret observations of drawings, pictures and schematizations from books
Listen to lecture on the embryonal development of the chicken
Observe tadpoles and caterpillars during the process of metamorphosis
Write individual reports and create drawings
*** The blood circulation**
Listen to lecture about the structure of the heart and its function
Discuss how to act out what happens inside the body to make it understandable to other children
Perform the play
Written individual reports about the travel of the blood inside the body
Revise and discuss the reports
Listen to lecture about the exchanges occurring during blood circulation

The sociocognitive interactions occurring in the small teams at work were analyzed separately; they have been compared across teams and across activities. Their description and interpretation is reported elsewhere [5].

Results

Contexts of Activity and Knowledge

Outcomes produced by the children in three of the activities they engaged in during the school year are described to give the reader some background to the major claims resulting from our analyses.

One task involved the children in constructing the insides of a rabbit (Table 2). Before completing this task, the children had written down their ideas about what makes up a rabbit. The written assignment had elicited mainly the external features of a rabbit. In contrast, during the construction task, children produced four models representing the inside of a rabbit. This activity was preliminary to the dissection of a rabbit and enabled the children to make their predictions explicit.

Table 2. Building a model of the inside of a rabbit

Assignment	Visualize the different organs that are inside a rabbit's body
Materials	A pre-arranged animal shape made with two interconnected empty boxes, one for the body and one for the head; different types of waste materials, such as wires, pieces of fabric, sponges, pipes, cardboard; tools.
Time	Approximately 2 hours
Social Setting	Group-work. The groups were composed of 4–5 children. The teacher and researchers supervised the activity, providing technical assistance, asking for explanations about what was being produced, and eventually replying to questions.

The novel aspects appearing in the models were:

- connections between the organs, at least in the alimentary, respiratory and circulatory systems;
- more inner organs: the bladder, the liver, the kidneys were included in at least three models;
- the presence of blood vessels as components of the inner organs, too, as in the stomach, the brain, the liver.

The qualities of the four teams' models were somewhat similar. There was only one where the intestine was represented as a coiled pipe longer than the size of the 'body-box'. During the experiment the children discussed:

- the analogical value of the objects and materials; this mainly occurred while children were selecting things in the container: their shape, color, and texture were evaluated making explicit reference to those assumed in the real parts of the body;
- the size and the proportions of the pieces representing the inside organs; this occurred when children were engaged in cutting, adapting and manipulating objects;
- the location and interconnection of organs; this occurred during the assembly of the pieces inside the cardboard boxes which simulated the body of the rabbit.

Essentially two types of strategies were adopted during the construction of the models: browsing through the available materials and picking up the ones that were evocative of certain parts of the body; and intentionally searching in the container for the most suitable materials, for those which best matched the design of the part that was going be realized.

The Assembly of the Poster

About two thirds of the school year had elapsed when the children were engaged in the task of explaining, through assembly of a concept map, "What is the organism and what does it do" (Table 3). This activity was preceded by a long sequence of work on the movement of materials. One of the concept maps was revised one month after its assembly.

Table 3. Assembling a concept map entitled "What is an organism and how it works"

Assignment	Assemble the labels on the white board so that other children watching the poster can get an idea of the relationships among the parts and the functions of the body
Materials:	A list of 60 words on a sheet of paper; a set of 60 cardboard labels having the same words written on them; some blank labels; a 100 x 70 cm white board; and pins. The list included the names of organs and functions, children's terms for naming various events and parts, and abstract terms such as: system, organism, energy, force, transport, life.
Time:	Approximately one and one-half hours
Social Setting	Group-work

Different solutions for organizing the labels were found by the four teams of pupils. The majority of the 60 labels were used: fewest were used by the group which imposed the highest degree of structure on them. It is interesting to note which of the labels were not used. Except for one team that used almost all the labels, the unused labels belonged to two categories: the more abstract and superordinate ones like 'organs', 'parts', 'organism', 'system', 'substances', 'transport' and some anatomical ones like 'esophagus', 'intestine', 'trachea', 'testicles', 'spermatozoa'.

One of the resultant maps was made up of many separate clusters of labels in sequence, each one containing a description of a part of the body. In another, the labels within each sequence were linked together in short strings with some short branches. In yet another map, serial strings of labels were afterwards interconnected to form a long sequence, but some shorter sequences were left isolated. Only one map displayed a hierarchical organization: three strings of words stemmed from the label 'brain', and each described one important function. This pattern had not been designed from the start: the children had first put together a long sequence of labels which described the leg and which included the brain; then they discussed the ideas that the brain controls everything and that the brain is an organism which gives orders to everything. Then they inserted the other sequences from the brain and began introducing what is in the brain. Their fellow students who were ultimately asked to interpret this map did not easily get its significance.

The maps and the conversations accompanying the activity suggest some general comments about the state of the children's conceptualization of organisms. The most complete set of labels of all deals with the leg. This reflects the great amount of time spent by the class on this topic. It reveals that better articulated knowledge was acquired, especially if we compare this team's performance with the individual reports written after the dissection of the rabbit leg. Some weak points still persist, though. Children used their own term 'wires' which has an uncertain reference in their discourse (sometimes it stands for nerves, sometimes for tendons). Furthermore, the functions of the nerves and tendons are confusing. In all the groups, children seemed to be aware of the role of the brain in the organism, but they gave more significance to its implications for movement and thinking. The relationships between the circulatory, the alimentary and respiratory systems do not appear in any form in the maps, in spite of the presence of the label with the words 'little doors' having been among the resources distributed to the groups. This term had been introduced by the children to identify ways through which food gets into the blood stream.

Children used different strategies for assembling the labels on the poster:

- browsing through the list of words which was suggestive of the topics, and only later making targeted searches for labels;
- looking for the necessary labels after having decided on a topic from the beginning;
- discovering associated pairs of words in the check list and then placing these pairs in the poster or merging them together.

The following kinds of social exchanges between the members of the work group appeared to improve the outcome of the activity:

- monitoring the work performed by a partner and eventually proposing labels that would fit the sequence,
- questioning peers on the relationships between labels,
- offering justifications for the identified relationships,
- commenting on and criticizing the order of words in the sequence while it was being assembled,
- going back to an assembled sequence to check if labels were missing,
- suggesting the substitution of the common language term by the scientific one,
- recalling observations, information and experiences relevant to making decisions,
- drawing attention to new topics essential for the map.

At the end of the activity all the posters were exhibited and discussed. The teacher asked children who had not been part of the group, but had assembled them to interpret the maps, to enable a shared assessment of their effectiveness in communicating the intended meaning to a reader.

The Play

The play was about what happens as blood travels along inside the body and was performed twice, first at the end of the 4th grade class, acted out by two teams only. The second performance occurred two months after the beginning of the second school year (in fifth grade) when all four teams played in it and were also videotaped (Table 4). This activity was performed two times, at the end of the 4th grade class and at the beginning of the 5th grade class. The second time a whole morning was spent preparing scripts and rehearsing. Each one of the four teams played the performance according to the prepared script and was videotaped. The audience could ask questions. A discussion followed and some days later the tape was reviewed and criticized.

The roles taken by the children during the first performance were: the blood, food, the lungs, and the foot, the latter as a very peripheral part of the body. In the second performance new roles were created for the brain and the liver. Explanations had to accompany actions on the gigantic body outline to enable other children to understand the story of the functioning of the body when watching the tape.

The students' discourses mentioned for the first time that the blood carries away 'dirty things' from the foot. They mentioned the textbook representation in which venous and arterial blood were distinguished by blue and red. Now they had acquired their symbolic meaning which had remained ambiguous and unexplained in the previous discussions. Another important introduction was the recognition that many different operations occur at the same time inside the body: "as the blood goes along, it unloads...," "as it passes by, some of the wagons enter the little doors in the food-pipe...," "perhaps there are many trains, one goes to one arm, another to the other arm, etc.," "Here we have one train, but it is implicit that there are many of them and, while one leaves from here, there is another one traveling past in the leg," "...now the food is going to...but in the meantime...," "E. loaded the air in only once, but actually she should have done that continually," and "...it's like when you have a pipe with many holes and you open the tap and the water spurts out of all the holes at the same time."

This kind of awareness appeared to have increased when children performed the drama in November. By acting out the metaphor of the course of travel of the blood-train departing from and arriving at different stations, the image of the double circulation and of the four partitions of the heart which had been previously illustrated by the teacher was consolidated. This acquisition could be found in the great majority of the children's writing which related the story of blood's path.

The situation enabled the teacher to raise stimulating issues which in that context could not and were not dismissed by the children because they felt they were significant: "If all the trains of the blood leave some stuff in the foot, then what happens to them in the foot?", "Where does the 'dirty blood' go after it leaves the foot?", "Can the blood take the food anywhere from the food-pipe?". And the children also identified problems: "But, do little doors actually exist (in the intestine or in the vessel wall)?", "What is it that makes the heart go bum-bum?", "I don't understand how the 'dirt' originates?", "Is the 'dirt' produced until we die? And does it increase after we die?"

Table 4. Simulating with drama what happens inside our body

Assignment	Take roles and play a drama to show to other children what you have learned about the functioning of the body, and what goes on in its inside when the blood circulates
Materials:	A gigantic outline (about 5 m) of the human body had been drawn on the floor by stitching a white tape. It contained the schematization of the alimentary and respiratory system. Colored cardboard cut into small pieces of different shapes and woolen threads were available. Four trains built by joining plastic boxes were used to represent blood flow. The heart with its partitions was schematized on a sheet of paper and placed over the body' outline.
Time:	Approximately 2 hours
Social Setting:	Two groups of four children were selected by the teacher to perform the drama, while all the school mates could add their comments and make proposals.

Histories of Conceptual Growth

When interpreting large protocols of ethnographic data, it is problematic to extract what might constitute evidence from them. How does the researcher make relevant to the reader the bits and pieces which acquire their meaning only when they are related to the entire flow of discourse, actions, events? Single statements are likely to be under- or over-valued but, on the other hand, over-detailed descriptions do not improve comprehension and may result only in boring the reader. Being aware of the risk, I will report a series of children's statements produced during the school year. They all deal with respiration and exemplify the existence of three explanatory models and the fleeting appearance of new ideas in different situations, which eventually drove the emergence of a fourth, more differentiated model.

The questions, "Why do we breathe?," "What is the air for?," and "What happens to the air when it gets inside?," were the topics of a long class discussion which the teacher led when the class was halfway through the curriculum.

> "The air enters the mouth, then we send it to the lungs and then we send it back."
> "A little of it remains inside."
> "It is to breathe, otherwise you die."
> "It is as if we eat it."

This is one of the models and half the class agreed with this 'depository view' of breathing in and out. One girl was to use a correct notion about air later on in the discussion but she essentially expressed the same model, although introducing the idea of selection: in the air there is oxygen and carbon dioxide, the oxygen is taken in and carbon dioxide is given off. This is another model:

> "The air first goes round ... first it goes into the lungs, then it goes into the blood and then you send it back outside."
> "Some of it remains everywhere." (Daniele)

Circulation of the air is mediated by the blood and therefore the transfer to the blood is imagined as possible; the other half of the class seems to be committed to this view. Stimulated by the teacher who presses the children to elaborate their ideas further, some of the children bring in a new argument: there is a difference between the air that goes in and the air that comes out.

> "The air which is inside is polluted." (Andrea)
> "If you don't give it off soon you get polluted too." (Emanuela)
> "If there is no air ...it may be that the sugar cannot be deposited somewhere." (Emanuela)
> "Without the air the blood cannot keep going." (Emanuela)
> "I cannot make out why the air is absolutely necessary!!!" (Emanuela)
> "The air goes inside, but when it comes out again it's different... though we cannot see it!" (Alex)
> "From the body the breath comes out." (Pino)
> "It contains air and water." (Andrea)
> "It has heat."

A third model seems to be hinted at by the statements which follow. A transformation operated by the body is the crucial, novel aspect to it:

> " It is polluted, though...the air which is close to the blood, this is why we breathe and that air pushes the polluted air in the blood and the clean air arrives."
> "... the air which is in the blood gets polluted ..."
> "... you have to keep it in order for the change [to occur]!" (Daniele)
> "When you give it off it is stuffy ... like ... 'breathed' ... inside the body or inside a room it has to be changed." (Sonia)

During the various activities and the verbal interactions which took place afterwards, these models interplayed. Some of their aspects came to the fore, triggered by the demands of the teacher or of the actual task, and some faded into the background. While interpreting the links between labels set up in the maps about the organism:

> "The brain breathes! ... it has oxygen." (Cristian)
> "The lungs make you breathe and take oxygen." (Maria)
> "A little bit of air is kept inside the lungs and drops into the blood ... the other part goes out." (Pino)
> "... on its way the oxygen ... it appears to get dirty ..." (Sonia)

After having observed invertebrates:

> "Not all tiny animals need air, but oxygen..." (Alex)
>
> "We take oxygen from the air...they take it from the water...Fish drink the water and then spit it back outside ...this water afterwards... takes oxygen again since there is air outside." (Carlo)
>
> "... the gills are like our mouth : they give off water." (Pino)
> "... we do not give off the same air that we breathe in, we put back outside the air that was already inside our body." (Carlo, Eleanor, Federica)
> "As when we open a window so that the stuffy air goes out and the good air enters." (Alex)
> "... the oxygen enters and does its tour, first it passes into the blood, then it goes all the way round and goes out, then it takes the new one." (Sabrina)
> "I would not write here: air plus oxygen, because the air is oxygen." (Carlo)

While rehearsing the play:

> "... the blood becomes blue, as Eleanor said ... and then it gets cleaned; ... [the dirty blood] goes to poopoo." (Cristian)
> "I [the blood-train] go and get air in the lungs, I carry it all around ... oxygen is needed; ... [the foot] gives back to the blood the dirty things of the air ... carbon dioxide." (Luisa)

While discussing the first performance of the drama:

> "...[every bit of the body] gives back to the blood the bad air...which is black...it is dirty blood...it already runs that way." (Fulvio)

While the teacher is summarizing the work done and lecturing about transformation processes and exchanges between the blood and the cells in every part of the body:

> "The blood takes the oxygen from the lungs but it also gives a little blood." (Eleanor)
> "... it may be that the [cells] that you have drawn on the board take in a tiny bit of oxygen...so that they take up nutrients and become substances of the body." (Luisa)

While each group is performing the drama, moving around the train of the blood:

> "... there are many cells which take nutrients from the blood since the blood had first taken food and air." (Eleanor, Federica, Giuseppe, Emanuela, Fulvio)
> "... the air arrives in the lungs and some air is given off," "the blood transports the good air around. Daniele puts in the bad air and it carries it back [to the lungs]." (Andrea, Luisa, Simona M., Simona N., Daniele)
> "... it travels up the foot and takes the bad food and the bad air...the blood enters the lung, it unloads the bad air and loads the good air, then it goes out of one lung and reaches the other one where it unloads a little of bad air which will be cleaned afterwards and takes some good air." (Maria, Pino, Cristian, Pamela, Moira)
> "... the foot gives it the old stuff ...it goes to the lungs and it takes some oxygen and it leaves some carbon dioxide." (Sabrina, Carlo, Sonia, Giulia, Alex)

After having revised the videotape of the performance:

> "In a machine there is the power that makes it go, but what gives the power to move our hand or our arm?" (Carlo)
> "In the body there is something like a battery, it is not connected to the power but is similar to a battery." (Maria)
> "I have the answer!! ... is the food and the water!" (Daniele)
> "And does the air get to do with this?" (the teacher)
> "Not much though ... the air does not feed." (Daniele)
> "... the blood carries to the kidneys the food that it did not succeed in transforming and from there it is sent to the wee wee." (Maria, Fulvio)

At this point a new model of explanation is characterized by some kind of awareness that the blood transports the waste it collects along its travels back to the lungs. Breathing out means to get rid of what originates inside the body and not of the 'bad' component of the air.

Thinking Strategies and Teaching

The conclusions that we drew from this kind of data have implications not only for the study of the processes underlying the organization of concepts, but also for the design of teaching methods which can effectively fit such processes. As exemplified here, each task mobilizes different ideas and constrains reasoning in specific ways. The construction of the inside of the rabbit bore on the spatial relationships between the organs, their proportions, number, continuities and discontinuities. It forced the retrieval of the fuzzy mental images, and of pictures and movies, that had to be brought to bear to solve the problem. It did not raise many conflicts among the children, probably because nobody was seriously committed to their own ideas. The assembly of the map from a given set of labels obliged the children to take into account all the sets of organs and functions and to question the meaning that could be assigned to the relationships, not necessarily only the spatial ones. The drama compelled the children to take time into account and enabled them to voice the multiple overlapping functions of organs. It also enabled them to visualize the relationship between the center and the periphery which make the flow of the substances through the body possible .

The negotiation of meaning among the children was supported by the sharing of a common goal (to produce a model) and by the manipulation of symbolic objects. That they *stood* for something else had to be agreed upon and this then meant that the children had to revise their mental models. The cognitive demands implied in the use of the symbolic media and tools were different in the various tasks. This allowed scope for various abilities to be required and for different children to become more central participants in the activities.

Each of the metaphors used for modeling structural and functional relationships stimulated children's comments at a meta-level of reflection. That is, on the representational value of the analogical devices and stereotypes for effectively conveying and communicating to other people what had been observed or understood of the real facts.

The discussions which followed the performances elicited a very active participation and the role that the teacher played on these occasions was very critical: he had to strengthen the shaky acquisitions that the construction of the models had fostered. The end-products were always illustrated by the children who questioned ideas, compared the facts, noted inconsistencies, recalled personal experiences and previous activities done in the classroom, and expressed new concerns. The teacher's interventions pointed to emerging topics or facts, underlined understated meanings, raised problems, asked different children to put into their own words what had been previously stated, made up a whole story from the various fragments, and made explicit which issues needed further comprehension and work. He sometimes translated into other representations (sketches, schematizations, outlines, etc.) the important ideas which were taking shape in children's discourses to fix them in a synthesis and, very importantly, he only slightly assuaged reasoning or ideas so that they would account more adequately for reality.

Some important cognitive obstacles were still present at the end of the school year. The children were still confusing two levels of understanding: the selection between good and bad things which occur at the level of the intestinal exchanges with the blood, and the production of waste substances occurring at the cellular level. Also the heart was somehow conceived as responsible for the 'purification' of the blood. But one important obstacle to the conceptualization of the organism had been removed. By inventing the presence of 'little doors' as a plausible approximation to account for the exchanges between the main body systems, the problem of system isolation was overcome. Children were aware that the doors were a tool of their scientific imagination and that reality was more complicated, but it was enough for their present needs. The exchange of substances at the periphery of the body, at the level of each 'tiny bit', gave meaning to the continuous flow of the blood and enabled the children to take a different perspective in thinking about respiration.

Not all the children progressed in the same way, but when a class works as a community of learners, as is suggested also by other contributors in this volume, the inputs of those students who reach a higher level of conceptualization are a potential resource for all the other children in the cognitive exchanges occurring during the negotiation of ideas, plans of action and, symbolization.

A New Approach for Modeling Conceptual Change?

The evidence from our protocols speaks in favour of a slow amplification of children's own repertoires of ways of thinking about the phenomena of the living body rather than of a restructuring of a conceptual network in which old ways of thinking are replaced with new ways of thinking. Many children's propositions about the meaning of the term organism did not change dramatically at the end of the school year. The same child who wrote "It's only a part of us, it's what we have inside," raised this insightful philosophical question in the context of a discussion to summarize the claims about the functions of the body: "I would like to put a question to all of you: if we are now trying to figure out what we have inside the body with our brain, which is inside the body, how is that the brain does not already know what we have inside the body?".

The rational process inferred a posteriori from the construction of scientific knowledge by children, and the succession of logical steps they took from one structure to another should help in modeling students' learning. The children learn in a situation which differs dramatically from the one in which professional scientists work. The effort of making sense out of reality is certainly the same, but the degree of awareness about the goals and the tools available is very diverse. Rather than disequilibration produced by cognitive conflict as a consequence of confrontations with anomalous facts, the mechanism for conceptual growth seems to be participation in practices which are congruent with 'learned' perspectives. Demand for consistency in the arguments grows and the need for representations that are reasonably faithful to the facts that they are accounting for increases. Gradually children begin to comprehend the purpose of using specific terminology, develop the ability to generalize from man to animals and vice-versa, and also the ability to observe and interpret evidence. All these components dynamically define a space, a system, where new analogies become necessary, new questions take form, and spatial and temporal relationships become possible in a world of scientific imagination which can also be counter-intuitive.

References

1. Arnaudin, M. W., & Mintzes, J. J.: Students' alternative conceptions of the human circulatory system. A cross-age study. Science Education, 69 (5), 721-733, 1985.
2. Bibace R. and Walsh, M. E. (eds.): New directions for childhood development: children's conceptions of health, illness and bodily functions. San Francisco, CA: Jossey Bass 1981.
3. Caravita, S., & Tonucci, F.: How children know biological structure-function relationships. Proc. Ithaca,, pp 64-73. Ithaca: Cornell University 1987.
4. Caravita, S., Tonucci, F., Consoli, V., Giuliani, G., & Rusca, G.: Children's reasoning about biological structures. In: Learning and Instruction (H. Mandl, E. De Corte, N. Bennett, & H. F. Friedrich eds.), Vol 2, pp 649-670. Oxford: Pergamon Press 1989.
5. Caravita, S.: La dinamica tra azione, discorso e prodotto nel lavoro di gruppo a scuola. In: Interazione sociale, mediazione culturale e conoscenza (C. Pontecorvo, ed.). Firenze: La Nuova Italia (in press).
6. Carey, S.: Conceptual change in childhood . Cambridge, MA: MIT Press 1985.
7. Carey, S.: Reorganization of knowledge in the course of acquisition. In: Ontogeny, phylogeny, and historical development (S. Strauss, ed.), Norwood, NJ: Ablex Publishing Corporation 1988.
8. Carey, S. and Gelman, R.: The epigenesis of mind: essays on biology and cognition. Hillsdale, NJ: Lawrence Erlbaum 1991.
9. Contento, I.: Children's thinking about food and lating: a Piagetian based study. Journal of Nutrition Education, 13, 86-90, 1981.
10. Fox Keller, E.: A feeling for the organism . New York, NY: W. H. Freeman and Co. 1983.
11. Gellert, E.: Children's conceptions of the content and functions of the human body. Genetic Psychological Monographs, 65, 291-411, 1962.
12. Gelman, R., Spelke, E., & Meck, E.: What preschoolers know about animate and inanimate objects. In: The acquisition of symbolic stills (D. Rogers, J., Sloboda, eds.). New York, NY: Plenum Press 1983
13. Gelman, R.: First principles organize attention to and learning about relevant data: number and the animate-inanimate distinction as examples . Cognitive Science, 14, 79-106 , 1990.
14. Gelman, S. A.: Children's expectations concerning natural kind categories. Human Development, 31(1), 28-34, 1988.
15. Gelman, S. A.: Children's use of categories to guide biological inferences. Human Development, 32 (2) 65-71, 1989.
16. Giordan, A., & De Vecchi, R.: Les origines du savoir . Neuchatel: Delachaux & Niestlé 1987.
17. Hatano, G., & Inagaki, K.: Everyday biology and school biology: how do they interact?
Quarterly Newsletter of the Laboratory of Comparative Human Cognition, 9, 120-128, 1987.
18. Keil, F. C.: Concepts, kinds and cognitive development. Cambridge, MA: MIT Press 1989.
19. Keil, F. C.: On the development of biologically specific beliefs: the case for inheritance. Child Development, 60, 637-648, 1989.
20. Inagaki, K.: Developmental shift in biological inference processes: from similarity-based to category-based attribution. Human Development, 32(2), 79-87, 1989.

21. Medin, D. L.: Concepts and conceptual structure. American Psychologist, 44, 1469-1481, 1989.
22. Posner, G. J., Strike, K. A., Hewson, P. W. and gertzog, W. A.: Accommodation of a scientific conception: toward a theory of conceptual change. Science Education, 66 (2) 211-227, 1982.
23. Strike, K. A. and Posner, G. J.: A revisionist theory of conceptual change. In: Phylosophy of science. Cognitive science and educational theory and practice (R. Duschl, R. Hamilton, eds.). Albany, NY: SUNY Press (in press).
24. Wellman, H. M., & Johnson, C. N.: Children's understanding of food and its functions: a preliminary study of the development of concepts of nutrition. Journal of Applied Developmental Psychology, 3, 135-148 , 1982.

Working with Personal Knowledge in Biology Classrooms on the Theme of Regulation and Homeostasis in Living Systems

Joop Buddingh

Centre for Science and Mathematics Education
Department of Biology Education and Institute of Education
Utrecht University, Princetonplein 5, 3584 CC Utrecht, The Netherlands

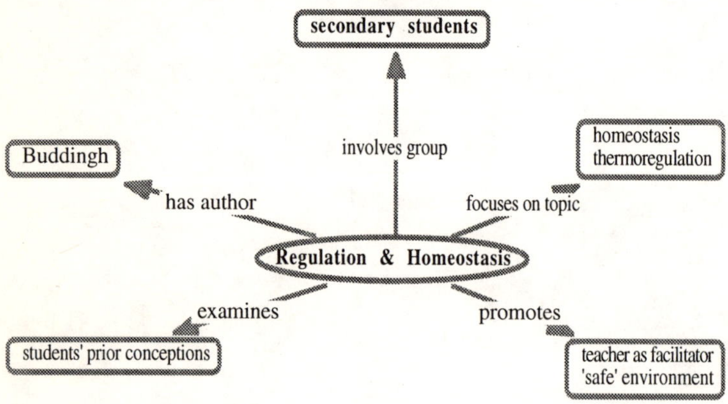

Abstract. This paper describes aims, methods, results and perspectives of an experimental project that can be characterized as engineering in biology classrooms. The paper starts with a description of the main issues used for designing educational scripts and materials, focusing on the philosophy of active exploration and (re)construction of knowledge by the students themselves. Then follows a description of the biological subject matter: content as well as relevance of this content will be placed in perspective of the foregoing ideas on education. We will then report on the classroom materials that have been developed, including the strategies to be followed in classroom practice. Finally research procedures, results and implications for teaching will be described.

Keywords. Educational research and development, biology education, biology curriculum, conceptual change, knowledge (re)construction, living systems, biological regulation, homeostasis, thermoregulation, human biology

Introduction

Biology education in secondary schools in The Netherlands will change in the mid-1990s. At that time the national final examination for biology will be based on a revised and partly renewed program. One of the themes which is more strongly emphasized, and therefore is to be renewed, is biological regulation with a special focus on the key concept of homeostasis. Neither tradition nor theory seem to exist to support biology education about this concept. Only a few studies mention difficulties in learning it [3, 21, 24]. The aim is to develop educational ideas and materials on this theme, supported and fed by a systematic research program. The outcomes of the project are intended for use by biology teachers and students at the pre-university level, aged 16–18 years.

Personal Knowledge and Shared Knowledge

Nowadays it is generally acknowledged that what people already know about themselves and their surrounding world plays an important role in the process of learning [9, 13, 18, 19, 20]. In a constructivist view on how knowledge is obtained, the prior knowledge of the learner fulfills a crucial role as a starting point and a bridgehead for new knowledge. Learning is considered to be a continuous process of reconstruction and expanding of prior knowledge. This understanding of objects and events becomes part of the learners' personal mental equipment. This may be called personal knowledge. When people communicate in a meaningful way about a specific object or event they understand each other. This means that the personal knowledge of an individual fits to the knowledge of other individuals. We call this shared knowledge. In the scientific world conferences and meetings serve as 'playgrounds' to negotiate ideas about the interpretation and understanding of the world. When a consensus in views and conclusions is reached, new elements are added to the shared knowledge of a specific discipline. Sharing of knowledge depends on the use of a system of common words and symbols: every discipline develops its own (written and spoken) language. Biologists for example use 'biologish' as disciplinary language, with its specific vocabulary and idiom. So, when personal biology knowledge can be communicated to other people within the community it becomes meaningful.

Working with Personal Knowledge

Classrooms and classroom groups can be considered as communities of (young) people who are trying to develop vocabularies for speaking about specific objects or events. In a biology classroom these objects or events refer to the living world. As a consequence, one of the major roles of the teacher is to create situations in which the learners are stimulated to make explicit their personal knowledge and to exchange their ideas with the other members of the group. In such settings it may be expected that learners gain more and more shared knowledge. Accordingly, the question arises as to what extent such personal knowledge can be elevated to the generally accepted level of shared knowledge – that is, to an understanding of biology that fits the understanding of biologists. This question relates to the aims of biology education and science education in secondary schools. It must be kept in mind that raising this question means that we are referring to a body of disciplinary knowledge, shared by an external group of people. If we want to avoid tension or even contradiction with a constructivist view on teaching it is necessary to state that essentials of the knowledge shared by scientific biologists should be used only as a reference and not as the only standard to judge a learner's understanding of biology! For classroom practice it suggests that students are developing a common language on essential features of biological objects and events, using suitable elements from the disciplinary language as an aid and a reference.

The Concept of 'Homeostasis' in Scientific Biology

To identify the essential elements of knowledge shared by biologists, we have made a broad survey of literature on the status and structure of the concept of homeostasis, as it is represented in scientific biological literature [5]. The results of this survey can be reported here only briefly.

Leading scientists in biology, like Ayala [2], Von Bertalanffy [4], H. W. Smith [22] and Prosser [23], qualify homeostasis as one of the basic properties of life: development and perpetuation of life on earth was and will be secured by regulation and self maintenance, frequently based on homeostasis. This points to the key function of this concept: the principle of homeostasis represents a quality that is unique and specific to living beings. For this reason alone the principle can be considered essential in biology education.

All studies concerning the origins of the concept point to the close relationship between the naming and elaboration of homeostasis by Cannon in about 1930, and the ideas of Bernard seventy years earlier [1, 6, 7, 10 11, 12, 14, 15, 16]. Bernard postulated, about 1860, that "La fixité du milieu intérieur est la condition de la vie libre", thus indicating the crucial role to be fulfilled by the cell's environment. Only within very narrow physicochemical limits can the survival of single cells be secured. In other words, Bernard related the constant qualities of body fluids, like blood and tissue fluid, to the capacity of animals to live a 'free life'. The properties of body fluids are decisive for survival of the organism. Bernard's ideas had a great influence on research into internal regulation mechanisms, at first in human beings and later in other warm-blooded animals. In Cannon's (1932) book *The Wisdom of the Body* he describes homeostasis as follows: "The constant conditions which are maintained in the body might be termed equilibria. That word, however, has come to have fairly exact meaning as applied to relatively simple physico-chemical states, in closed systems, where known forces are balanced. The coordinated physiological processes which maintain most of the steady states in the organism are so complex and so peculiar to living beings – involving, as they may, the brain and nerves, the heart, lungs, kidneys and spleen, all working cooperatively – that I have suggested a special designation for these states, homeostasis. The word does not imply something set and immobile, a stagnation. It means a condition – a condition which may vary, but which is relatively constant." Cannon discusses some examples of homeostasis in the human body, but also speaks about homeostasis in general, using the word as a collective noun, i.e., an indication of the summative effect of all these equilibria. In this elaboration of the concept, the idea of stabilizing internal mechanisms gives the original idea of Bernard new dimensions.

After Cannon launched his ideas, the concept spread to various scientific disciplines. As a result of our study we conclude that in most of the sciences (except the life sciences) the word is used merely as a synonym for equilibrium, balance, stability, etc., without a clear disciplinary elaboration. In these other disciplines the concept has vague meaning and is often used as a metaphor.

In life sciences the concept obtained its prominent position from the physiological disciplines. These disciplines provided its empirical scientific basis by supplying a large amount of experimental data about regulation processes. General systems theory and cybernetics clarified the understanding of the phenomenon by providing models and representations suited to describe and explain homeostatic mechanisms. As an example we mention the close connection between these mechanisms and the capacity of self-maintenance by organisms, characterized by a continuous exchange of matter, energy and information with their environment.

In brief, the concept of homeostasis refers both to the result and to the processes of regulation by means of feedback mechanisms with a set point incorporated in the system. These types of regulation proceed on the levels of cells and organisms, that is, in autopoietic systems of the first and second order [17].

For clear communication about the concept it seems useful to distinguish between homeostatic regulation and homeostasis, following Barass [3]: Essential to understanding the mechanisms and function of homeostasis is insight into the relation between homeostatic regulation and the interaction of a system with its environment.

Classroom Materials

In close cooperation with a small group of biology teachers a set of student texts has been developed with the working title *Surviving by Regulation*. Main issues are the relations of living systems with their environment, illustrated at the levels of cells, organisms and populations, and dealing with regulation at each of these levels. The concept of homeostasis is elaborated at the level of organisms using the example of thermoregulation in man. The material comprises a mixture of information and tasks that have to be elaborated firstly individually, and then in small groups. To design tasks that give students the opportunity to explicate personal knowledge, a series of pilot studies has been carried out in classrooms. The aim of these studies was to explore procedures, contents and contexts suitable both for use in regular classroom practice, and for collecting research data without interfering with normal classroom practice. The next example illustrates this in more detail.

With respect to thermoregulation in man, students start with a task similar to the CES task designed by Simpson & Marek [21]. This task describes what happens when a person is subjected to heavy physical exercise. Students were to use their personal knowledge about these phenomena. They were asked first to explain individually why respiration and heartbeat frequency change, and why body temperature remains constant, and to design a scheme illustrating this regulation. After having made up their minds individually on this task, students discussed their views on the problem in small groups. As an outcome of their discourse each group presents a scheme on temperature regulation. After the exercise students were confronted with information on regulation of body temperature and on the main organs and processes involved. In the next task student's personal knowledge about fever was explored. Students were invited to give their own definition of the concept of fever. Then a text is given on the illness of a person, named P, describing changes in his body temperature, including a strong rise, a crisis and a rapid decrease within 48 hours. The course of the fever was also depicted graphically. In the following task students have to explain what exactly happens during the illness, and to compare all these facts with the information (including the scheme) on the temperature-regulation. This task culminates in questions on the role of the temperature-regulation system during the illness of person P. Finally the student is asked to rethink, comment upon and eventually reformulate his former definition of fever.

Research Procedures

An important aspect of this project is the close relation between the development of materials for use in the classroom and the research activities. Researchers worked in close connection with normal classroom practice and avoided interfering with classroom procedures as much as possible. For this reason data about task performance by the students were collected from work sheets produced by individual students and groups in the normal course of classroom events.

Individual and collective learning effects were followed over a period of about four months. Before starting with the classes, students were pre-tested by means of paper and pencil tasks containing problems and questions similar to those in the classroom materials.

Immediately upon completion, students were post-tested with the same questionnaire. To identify retention effects, all students were asked about three months afterwards to answer this questionnaire again. An analysis of the data on every individual, collected on four occasions, enabled the project to follow trends in learning effects. An important aim of this analysis was to assess an eventual shift from heterogeneous personal knowledge to more homogenous shared knowledge. Also collected was data about the student's evaluation of the lessons and the materials used. To measure learning effects a traditional product-evaluation, with open questions, was performed immediately after the end of the series of lessons and apart from other research activities. These data were also to be taken into consideration in an overall evaluation of the project.

Results, Observations and Comments

We first report on learning effects and conceptual change, focusing on one example of homeostatic regulation, thermoregulation in humans. As mentioned before, the design of the lessons on thermoregulation was preceded by a series of pilot studies. As a result of these studies we found that student ideas at ages 16–18 can be classified in two ways.

In the first, about 20% of the students took the constancy of their body temperature for granted: 'a constant body temperature is due to the body.' Some typical examples of these explanations are:

- The body cares for a constant body temperature,
- You are trying to keep your body temperature as constant as possible,
- The body tries to keep its temperature at 37°C,
- The body tries to avoid changes in body temperature,
- Man is warm-blooded: the body temperature has to be kept at 37°C,
- The temperature inside the body has to be constant.

These students refer in their explanations to a disruption of this constancy as illness. A change in body temperature is pathological as in:

- Body temperature only changes if you are ill,
- A non-constant body temperature is unhealthy,
- After heavy physical exercise you don't get a fever.

Most of the others (about 80%) see a direct link between the constancy of their body temperature and sweating. They think that 'sweating regulates body temperature.'

Their explanations can be subdivided into two types:

- Body temperature doesn't change as a result of sweating, or
- Body temperature changes, this *causes* sweating, followed by a fall of body temperature.

What's going on in the inside of the body, the processing of warm and cold, seems to be for all students something magic!

Student ideas on thermoregulation are based on a black-box view on the human body. Translated into goals of biological education, one of the aims could be to help the students to transfer from a black-box relation with their own body to a glass box relation, i.e., teach them to look inside their own body.

This view was applied to the design of the series of lessons: the method used in the pilot studies seemed practical and useful to help students explicate their ideas and explanations about maintaining a constant body temperature. Students were asked to explain why body temperature does not change when a person engages in heavy exercise.

The student answers, collected prior to the classroom lessons, are very similar to those in the pilot studies; the two groups of students used the same types of explanations. Obviously students of this age share the same ideas, rooted in a black-box perspective on their own body. This seems to be a widespread and persistent element of a student's personal knowledge. After having explicated their ideas on maintaining a constant body temperature, students studied a text dealing with the structures and processes involved in temperature regulation in humans. Data were collected three more times during and after the lessons in order to follow trends in student thinking and reasoning about temperature regulation. The collected data gave insights into learning effects over a period of about four months.

An important part of the unit on thermoregulation consisted of a task on fever. Here again the students were asked to first formulate their own views and opinions on this phenomenon as a starting point, followed by steps of discussion, reflection and an invitation to rethink their ideas. To record what happened with student concepts of fever we identified keywords in the student explanations, like 'body temperature', 'set point', 'caused by', 'disrupted', 'raises', 'falls', etc.

In their first explanations all students said that fever is synonymous with illness and a rise in body temperature. They did not apply the previous information on temperature regulation to the phenomenon of fever. But after elaboration of the task on fever, and discussions in small groups and in the classroom, we could observe that students became aware of the role of the temperature regulation system: they were using more and more the same language and ideas and, in other words, became able to share knowledge.

In a first posttest at the end of the series of lessons, a clear shift was observed in student explanations, from those that only referred to raised body temperature and/or illness to causal explanations, which showed insight into a change of the set-point caused by illness, allowing the body temperature to rise. In a second posttest, about three months after classroom practice about 80% of the students still used these causal explanations.

From other observations made during the lessons, we report the following. The way the teachers were guiding the students through the texts and tasks was very crucial. They encouraged them to come with their own ideas and views, and helped them to discuss and to negotiate their opinions. Generally the classroom climate could be characterized as permissive and stimulating. Secondly, students were highly motivated to communicate about their own, i.e., personal views and experiences with fellow students. This can be explained from two perspectives. The first is that it had to do with the permissive and stimulating classroom climate, implying the implicit and explicit message to the student: your experiences and ideas are important and worth communicating. The second is that students were working at that moment with the object they have had a life-long experience with, that is, their own body.

Implications for Teaching: Conclusions and Perspectives

After the experiences of working with personal knowledge in regular classroom practice, a first conclusion could be that working with such knowledge in this way seems practical. According to the teachers, the interference between their regular classroom practice and the research activities was minimal.

In this design the teacher has an important task to create conditions necessary to facilitate students' work with their personal knowledge – that is, to encourage students to use their own words and ideas to express their understanding of reality. This demands a permissive classroom climate. The teacher's role is to stimulate students to give their views and to promote discussions and negotiations about these views. So that personal student knowledge evolves into shared student knowledge, a decisive condition is that teachers promote social learning: talking, thinking and negotiating together.

An important question, worth exploring further, is how to combine the philosophy of social learning with the use of new instructional technologies and procedures. Can they be combined in a fruitful way in regular classroom situations?

This project illustrates the importance of exploration of personal knowledge for working on conceptual change in classrooms. The strategies we used were highly motivating for the students: as illustrated by the very vivid discussions and problems that arose. Students explored themes and questions of personal relevance. The following example illustrates this. After having discussed the effect of aspirin as a suppresser of fever, a girl asked: "What happens when I take an aspirin when my body temperature is normal? Does it fall?" Evidently she was able to pose and think about interesting new questions, demonstrating a kind of biological literacy as described by Demastes & Wandersee [8].

This example could be connected with a plea for designing instruction in which students explore their knowledge of their own body. Why? Because it is the most well-known biological object for everyone. Each individual builds up a gigantic, lifelong personal knowledge of his or her body. Students, too, have this knowledge and at the ages between 12 and 20 years old are particularly interested in their bodily functions and are highly motivated to communicate with one other about their bodily experiences.

This approach would also provide an opportunity to work on questions about personal and collective responsibility for well-being, that is, health education. It also illustrates that observations made with people can be generalized to other mammals as well as vice versa. Insights into the principles of the being, behaving and becoming of an organism provides a basic framework for structuring knowledge to allow an understanding of the unique properties of living beings and, more generally, of living systems. This seems to be a promising and fruitful way to transfer responsibility of the individual student for her or his own well-being to the well-being of other human beings, an important step toward their acceptance of responsibilities as citizens of planet Earth. In this way, working with relations and knowledge structures in biology classrooms could be connected with the ultimate aims of biology education.

References

1. Adolph, E.F.: Early concepts of physiological regulations. Physiol. Rev., 41, 737-770, 1961.
2. Ayala, F.J.: The autonomy of biology as a natural science. In: Biology, History and Natural Philosophy (A.D. Breck, W. Yourgrau, eds.). New York: Plenum Press 1972.
3. Barass, R.: Some misconceptions and misunderstandings perpetuated by teachers and textbooks of biology. J. Biol. Educ., 18, 201-205, 1985.
4. Bertalanffy, L. von: The model of open systems: beyond molecular biology. In: Biology, History and Natural Philosophy (A.D. Breck, W. Yourgrau, eds.). New York: Plenum Press 1972.
5. Buddingh, J.: The status of homeostasis. Utrecht: Centre for Science and Mathematics Education 1995 (in prep.).
6. Cannon, W.B.: Organization for physiological homeostasis. Physiological Rev., 9, 399-431, 1929.
7. Cannon, W.B.: The wisdom of the body. New York: Norton 1932.
8. Demastes, S., Wandersee, J.H.: Education: Biology literacy in a college biology classroom. BioScience, 42, 63-65, 1992.
9. Driver, R., Guesne, E., Tiberghien, A.: Children's idea's in science. Milton Keynes: Open University Press 1985.
10. Goodfield, G.J.: The growth of scientific physiology. London: Hutchinson 1960.
11. Hardy, R.N.: Homeostasis. London: Arnold 1983.
12. Holmes, F.L.: Joseph Barcroft and the fixity of the internal environment. Journal of History of Biol., 2, 89-122, 1969.
13. Jonassen, D.H.: Objectivism versus constructivism: do we need a new philosophical paradigm? Educ.Techn.Research.& Development, 39(3), 5-14, 1991.
14. Langley, L.L.: Homeostasis. London: Chapman & Hall 1966.
15. Langley, L.L.: Homeostasis: Origins of the concept. Stroudsburg (Penn.): Dowden, Hutchinson & Ross 1973.
16. Leake, C.D.: Perspectives of adaptation: historical backgrounds. In: Handbook of Physiology (D.B. Dill, ed.), Section 4: Adaptation to the environment. Washington D.C.: Am. Physiol. Soc. 1964.
17. Maturana, H.R. & Varela, F.J.: The tree of knowledge: the biological roots of human understanding. Boston: Shambala 1984.
18. Novak, J.D.: Learning science and the science of learning. Studies in Science Educ., 15, 77-101, 1988.
19. Osborne, R., Freyberg, P.: Learning in science. Auckland: Heinemann 1985.
20. Scott, P.: A constructivist view of learning and teaching in science. Children's Learning in Science Project. Leeds: Leeds University 1987.
21. Simpson, W.D., Marek, E.A.: Understandings and misconceptions of biology concepts held by students attending small high schools and students attending large high schools. Journal of Research in Science Teaching, 25, 361-374, 1988.
22. Smith, H.W.: From fish to philosopher. New York: Doubleday 1961.
23. Prosser, C.L.: Levels of biological organization and their physiological significance. In: Ideas in modern biology. (J.A. Moore, ed.) New York: Nat. Hist. Press 1975.
24. Westbrook, S.L., Marek, E.A.: A cross-age study of student understanding of the concept of homeostasis. J. Research in Science Teaching, 29, 51-61, 1992.

Generating Connections and Learning in Biology

Malka Gorodetsky[1] and Kathleen M. Fisher[2]

[1] Departments of Education and Chemistry, Ben Gurion University, Beer Sheva, Israel
[2] Center for Research in Mathematics and Science Education,
San Diego State University, San Diego, CA 92120-5006, USA

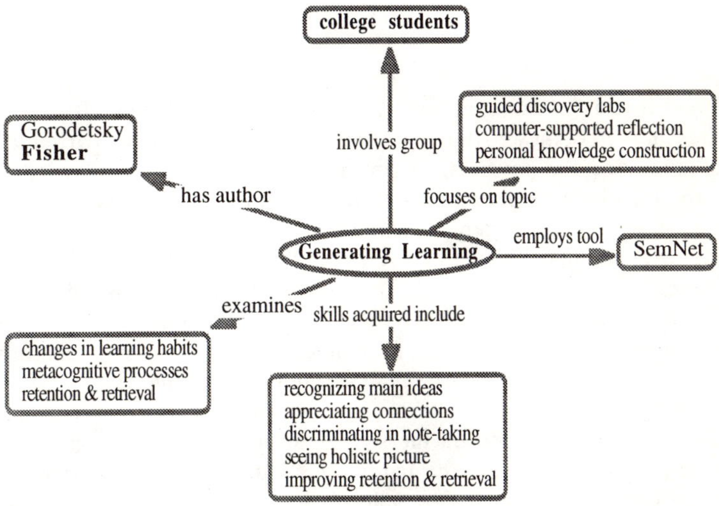

Abstract. In this chapter we examine the impact of teaching strategies designed to promote conceptual understanding among prospective elementary school teachers enrolled in a senior level college biology course. Instructional strategies include hands-on experiments and demonstration tasks designed to challenge naive conceptions. Activities utilize everyday materials so that they can be readily adapted to elementary classrooms. There is relatively little lecture, with strong emphasis on small group collaboration and whole class discussions. Students are prompted through questioning and example to develop runnable mental models of the topics being studied. These are comparable to the situation models proposed by Kinstch and van Dijk [18, 19] and discussed in this book by Otero (Chap. 3) and Scardemalia (Chap. 4). Students use a computer-based tool, SemNet™, to make their thinking explicit and visible, to reflect upon their understandings, and to share their thinking with their peers. The chapter describes studies to measure changes in student learning habits, metacognitive processes, retention and retrieval, and learning. SemNet students exhibited significant increases in deep processing. The volume of information retained and retrieved about a topic (the digestive system) by SemNet students was nearly twice that of the Comparison Group. SemNet students acquired certain cognitive skills (such as identifying main ideas and tying ideas together) that carried over into their other courses, according to students' self-reports. There is evidence for a low level of metacognition among SemNet students (that is, awareness about thinking processes is elicited with specific prompts but not generated spontaneously). Neither Group exhibited transfer skills.

Keywords. Biology, conceptual understanding, meaningful learning, deep processing, SemNet™ software

Context and Goals

The teaching and research described here are grounded in constructivism. Yet each teacher/researcher comes to the constructivist paradigm with particular background assumptions and perspectives; in our case, our views of knowledge construction are informed largely by cognitive science, linguistics, science education and artificial intelligence, combined with appreciation of the importance of conversation and social interaction in learning [8]. Thus, we might be described as 'cognitive constructivists'. This viewpoint should be useful to the reader, since each perspective has a contribution to make the total constructivist picture, but each also has its own language and ways of thinking, making effective communication across perspectives sometimes challenging.

In a simplified cognitive view, humans function in part by receiving information from the environment via the senses, storing selective portions of that information for later retrieval, and using stored information to anticipate events and make predictions. Further, people tend to perform best in familiar environments where they can draw upon vast stores of personal knowledge about those environments, although this is balanced by the heightened attention and awareness that often accompany immersion in unfamiliar environments.

Theoretically, the skills and strategies for making sense of situations that students develop over the course of their lives should serve them well in academia, even though academic learning makes demands that, in at least some ways, go beyond everyday life. In fact, however, a schism has developed such that 'making sense of the world' is often not perceived to be a major emphasis in science classes, especially among non-science majors. Rather, the large volumes of factual, word-based information to be learned, the relatively small number of opportunities for direct experience and authentic problem solving, and the widespread use of multiple choice testing have led many students to rely largely upon rehearsal learning.

For these students, academic learning involves a cycle of memorization, recitation, and forgetting. There is a sharp distinction between real-world meaning-making and such classroom science learning, with little motivation to create links between the two worlds of knowledge. Most of the students in this non-majors class share this world view. They have well-developed memorization strategies and initially resist our efforts to prompt a significant shift toward meaningful learning. As seniors in college, their cognitive habits are well established. However, after an initial period of anxiety, these students generally find that they like learning for understanding and, further, they feel the computer analyses they generate are constructive, not burdensome, tasks.

In this chapter we examine the impact of teaching strategies designed to promote conceptual understanding among prospective elementary school teachers enrolled in a senior level college biology course. The strategies include hands-on experiments and demonstration tasks designed to challenge naive conceptions and help develop scientific conceptions. The activities generally utilize everyday materials so that they can be readily adapted to elementary classrooms. The course includes relatively little lecture, with strong emphasis on small group collaboration, individual network construction and whole class discussion. Students are prompted through teacher questioning, questions posed in their laboratory manuals, problem-based assignments, and frequent quizzes and tests, to develop runnable mental models for each of the topics being studied. In addition, students use a computer-based tool, SemNet™, to make their thinking explicit and visible, to reflect upon their understandings, and to share their thinking with their peers. Thus, the course we describe here is familiar to students in many respects but unique in several ways. The course has been adapted from [2].

The emphasis during the semester described here is on personal, individual knowledge construction by students. We describe evidence of changes in student learning habits, metacognitive processes, retention and retrieval, and learning, primarily qualitative in nature. We also comment on changes made subsequent to this study, based on what we learned here.

Learning Biology Through Guided Discovery Labs with Computer-Supported Reflection

The biology course described here employs many of the principles of constructivist science teaching espoused by Tobin [9, 10, 11, 12, 20] and others. In addition, a unique feature is the full-scale integration of the SemNet™ software into the course. SemNet is used routinely by students throughout to make their knowledge of biology explicit, reflect upon it, and polish their conceptual organization. Pre-post comparisons within the course assess the impact of the entire instructional milieu, while cross-section comparisons evaluate the impact of SemNet.

Among the underlying constructivist assumptions guiding our work are that:

- Students often have their own personal theories of knowledge;
- Personal theories are sometimes robust and often implicit;
- On the other hand, many students simply acquire knowledge fragments which they don't even try to connect together, left to their own devices; this is the prevailing state of students' biology knowledge when they come into our course, as near as we can tell;
- Students' personal theories and/or knowledge fragments provide the starting point and foundation for successful instruction;

- Some personal theories, such as that 'plants do not respire' or that 'photosynthesis *is* respiration in plants', are resistant to change;
- Teachers can prompt students to give up their naive theories by careful selection of demonstration tasks to challenge key misconceptions, as when we grow plants sealed in plastic baggies for three weeks or more;
- Interpretation of such observations requires thoughtful discussion, analysis and reflection in a 'safe' environment in which students can share their doubts and alternative ideas without ridicule or penalty; when we employ 'safe' discussions in conjunction with the 'baggy experiment' and several others in which we measure CO_2 levels, for example, students come to accept that respiration and photosynthesis occur simultaneously and can be self-sustaining in a closed environment;
- Reflection can be promoted and supported with appropriate tools such as SemNet, and
- Students can generally think, reason, and understand far better than teachers initially believe.

Constructivist learning is slow at first, but there is evidence, especially from some of the mathematics projects, that it can be cumulative and powerful. Constructivism aims to promote deeper understanding and when understanding of an idea is deep, it needs to be learned only once. Stated another way, when students really understand something, they don't forget it. Our assignments prompt students to develop runnable mental models because such functional organizations appear to be more useful than inert knowledge. Once these coherent cognitive structures become part of students' mental landscapes, their world views, they can simply run their models and freeze-frame them at a particular point to answer a particular question. In addition, once ideas are comprehended reasonably well, they can be extended and applied with a fair degree of success to at least closely related new situations. When these kinds of changes are achieved, constructivist learning, in spite of the initial slowing down and paring of the curriculum that is required, can potentially accelerate student learning and thus also accelerate the curriculum, although we don't go that far here in examining a one-semester course.

SemNet enables the learner to construct networks of ideas by naming concepts and describing the connections among them [4]. Networks can be constructed individually or collaboratively. There is no practical limit to the size of the computer-based networks that can be produced with respect to either the number of concepts included or the number of connections among them. All concepts and connecting relations are generated by the users. Since nets are multidimensional and therefore difficult to see in their entirety, a net is typically examined on the computer screen one central concept at a time with all its links to related concepts, known as a graphic frame (Fig. 1). Many alternative views of the network are also available as described elsewhere [4, 5].

To obtain two-dimensional overview maps, students print graphic frames from SemNet and paste them together, often with appropriate drawings and illustrations [5]. The process of generating networks helps learners to identify gaps in

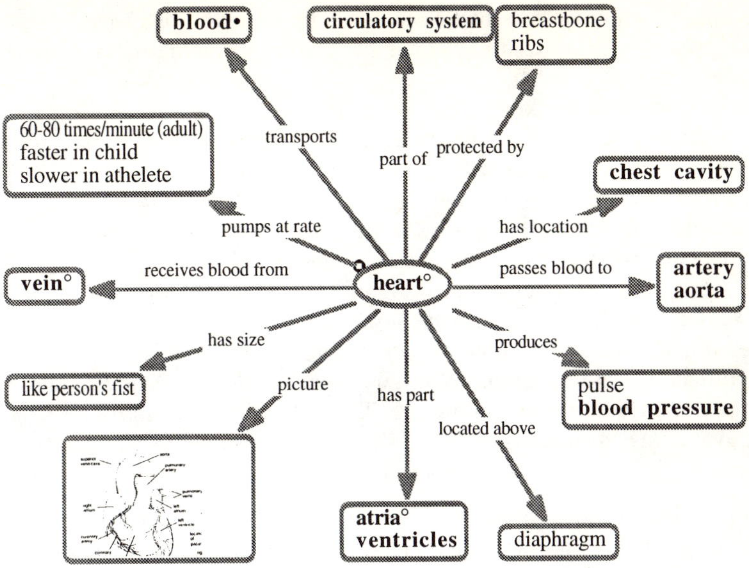

Figure 1. A graphic frame of a semantic network showing a central concept (heart) with its links to related concepts

their understanding and share and compare their ideas with peers. The computer screen serves as an extender for users' short term memories, allowing them to consider relations among many ideas at once. Since SemNet holds students' ideas in front of them in a coherent skeletal framework, they are able to use their working memory to reflect upon the appropriateness and adequacy of those ideas and the ways in which they have put them together. Some of the intrinsic features of SemNet that we find useful for promoting learning processes are the following.

(1) SemNet is an organizational tool that can be used by the student to summarize knowledge in a concise and precise manner, creating a skeletal framework.

(2) Elaboration and revision of networks is fairly easy and can be done at any time. There are mechanisms for connecting new knowledge to old and for editing and reorganizing knowledge representations.

(3) Connectiveness and linking are key features of SemNet and are also key features of understanding. User-defined relations are used to connect ideas together into multidimensional frameworks. In our college biology classes for prospective elementary school teachers, for example, student-generated representations have included 1–150 different kinds of relations linking together 30–2500 concepts.

(4) Language is the medium by which SemNet operates. The software requires that students name concepts and then establish named connections among those concepts. Actually it stimulates the student to explicitly describe

connections which otherwise may remain in a fuzzy and implicit form in the students' mind, helping to prompt their shift from perceptual to conceptual knowledge.

(5) SemNet is a tool for thinking. That is, the process of constructing a net is generally more valuable than the product that is created. The construction process is valuable in part because it:
- engages students in thinking about the subject being learned;
- requires them to transform their knowledge into new representations; and
- prompts them to make their implicit knowledge explicit, to create linkages among ideas, and to systematically organize their knowledge using such devices as categories, hierarchies and causal chains.

(6) SemNet mirrors students' thoughts, capturing them for leisurely examination, reflection, and revision. With this support, students can think about and reflect upon how they are thinking. This reflection can lead to "recognizing, evaluating, restructuring" [6]. Their networks serve as a summary of past learning as well as a starting point for broader and deeper constructions.

In addition to the potential learner benefits, we find that reviewing student nets provides insights to us, teacher/researchers who are interested in students' understandings. In each midterm and final and in some quizzes, students are asked to construct a network to describe a particular topic. Students typically work individually and from memory, without access to any resource material. This manner of testing provides a graphic picture of not only what ideas the students bring to bear, but also how those ideas are organized. Because networks require concise and precise 'skeletal representations' of knowledge, it is not possible to bluff or ramble as students often do in writing essays. Alternative conceptions, incomplete models, and erroneous propositions are readily apparent. We usually grade SemNet tests like essays, looking for such attributes as clarity, coherence, and completeness.

In this chapter we describe the impact of SemNet on prospective elementary school teachers in a capstone biology course in which SemNet is used as a learning tool. The course is a 'hands-on' class that aims to consolidate students' biological knowledge as well to serve as a model for teaching biology effectively in elementary and middle school. SemNet is used regularly by the students to summarize the biology topics they are reviewing. Student learning, cognitive style, and metacognition were monitored.

Methods

Two classes of the capstone biology course for prospective teachers were compared, one taught by the second author using the SemNet software and a second section taught by an experienced and popular teacher but without SemNet. The two classes covered many of the same subjects, but the instructors favored different emphases. Instruction in the SemNet Group was more conceptually and theoretically oriented, with emphasis on students' developing conceptual understanding of selected ideas in biology. Instruction in the Comparison Group was more process oriented, with emphasis on the development of skills for conducting, interpreting and reporting scientific experiments. Controlling variables, designing controls, and constructing graphs are examples of skills emphasized in the Comparison but not the SemNet class. Thus, while the data reported here are comparative they do not represent a controlled study, and the outcomes that are described should be interpreted accordingly. Both classes studied the digestive system which is evaluated here, and the two groups spent about the same amount of class time on the topic. The study was done using the following measures.

Measuring Changes in Learning Habits

Information regarding changes in learning habits and strategies was obtained through interviews with six students from the SemNet Group. The students were chosen to represent different performance levels in the class (2 high, 2 middle, 2 low). Interviews were conducted three times during the semester (beginning, middle, and end). The interview was unstructured and focused on habits of learning, study strategies, the use of SemNet, and other related subjects. Interviews were conducted by the first author and Dr. Barbara Wyman, who were not course instructors, and were transcribed and analyzed.

Schmeck's Inventory of Learning Processes [14, 15], a self-report survey, was administered to the two groups at the beginning and end of the course. The 62-item self-report questionnaire consists of four scales: Deep Processing (e.g., "I read critically"), Elaborative Processing (e.g., "New concepts usually make me think of similar concepts"), Methodical Study (e.g., "I have regular weekly review periods"), and Fact Retention (e.g., "I do well on examinations requiring factual information").

Measuring Metacognitive Processes

Quantitative and qualitative data regarding metacognition were collected by a task geared to reveal metacognitive processes while learning [13]. Students in both sections were asked what procedures for promoting learning they might suggest to a friend who is already a high achiever and to another friend who is a low achiever. These questions were asked twice, once at the beginning and once at the end of the course. The suggestions were analyzed into categories similar to those proposed by Salomon [13].

Measuring Retention and Retrieval

In order to assess the effect of studying with and without SemNet, both Groups were given an ungraded, three-question test at the end of the semester. Question 1 aimed to determine if the performance of the two Groups was comparable on a common assignment. Each Group was given a condensed article, "Frozen and Alive" [17], and asked to study it carefully and to take notes. They were told that they would turn both the article and their notes into the researcher when they were done studying it, after which they would be tested to see what they remembered about the article. In the test, students were asked to summarize the main ideas of the article in short propositions or statements.

Question 2 was designed to compare students' abilities to organize and retrieve knowledge attained by studying with pen and paper compared to knowledge attained by studying with SemNet. Students were asked to write a series of short, simple propositions describing the digestive system. Both classes had studied the digestive system during the semester, the SemNet Group with the aid of SemNet and the Comparison Group without SemNet.

Question 3 was designed to assess transfer of ideas from the recently-read "Frozen and Alive" article. The article described physiological adaptations of animals, particularly frogs, to extreme cold. In the third question, students were asked to described the long term effects of extreme heat on populations.

Questions 1 and 3 were graded on the quality of their biological content. Students in the two Groups had used similar, controlled study methods to acquire the information needed to answer these two questions. Results suggested that the two Groups were, on average, of comparable ability for these tasks, both with respect to the accuracy and extent of their knowledge. Students in the two Groups had studied in different ways for Question 2 (the digestive system). Their responses were analyzed both quantitatively, focusing on the number of propositions and the number of words each student generated in writing these descriptions, and qualitatively, involving judgments about content.

Results and Discussion

Fear of Science

Earlier studies revealed that students generally come into this class fearing and disliking science, and that their fear gradually gives way to enjoyment:

> I'm not a science buff, and if anything, scared of science classes because I do not do well in them, but I do try the very best I can do. (MM, 1/29/90)
> I enjoyed the symposium; I'm starting to question more and feel more confident. This class is becoming more fun every day. (MM, 4/2/90)
>
> I am intimidated by all science ... In the past science has seemed confusing, tedious, boring. (JR, 1/28/90)
> I really enjoyed the last lab we had on flowers and identification. It was exciting to find out the different parts and sexes of the flowers. I didn't even realize flowers had different sexes. (JR, 5/7/90)
>
> I always found science to be difficult and confusing. (CO, 2/5/90)
> I feel that because of this class, I am becoming more open-minded to science. (CO, 2/11/90)

Memorization Versus Exploration

As noted in the introduction, students in this class are practiced in memorizing. They often refer to their former dependence upon memorization and external authority:

> There is a big difference in memorizing facts and knowing facts. There is also a big difference between looking lab answers up in the book, summarizing or paraphrasing them ... and making the information really mean something to you. (CR, 3/5/90)

> It is refreshing to be able to visualize processes and understand them instead of just memorizing theories and parts of a molecule. (CH, 2/26/90)

> I am learning much more by using all my senses, visual, tactile and auditory. I am also retaining it for longer. I can memorize virtually anything to pass a test, but I find that once the test is over, so is my memory of the materials. (EH, 2/19/90)

> Once I have done a lab, I can usually picture what the results were. In contrast, if I read about what happened, I may not remember as much. (TK, 2/26/90)

> I find it very difficult to solve problems without using a book (or other authority) to refer to for support. It is much more stimulating and satisfying to solve something or invent something on my own; how did I get through college memorizing and repeating information from books and lectures? (TM, 2/5/90)

Changes in Learning Habits

In this context, we wanted to look closely at the changes that were occurring in students. Their initial learning strategies, as described in interviews by six students in the SemNet Group, have common features. All of them are linear, centered on the notes of the course with a strong orientation toward the final grade. Students expressed a strong desire to use familiar learning procedures, as explained by these students.

> Because I have done note taking that way for a long time and if I change the way I do it now then I get confused. (J, 6)

> And I've been passing you know, I haven't failed yet, maybe when I get my failure then I'll say, change your strategy. (M,7)

Students were expected to shift from their largely linear, compartmentalized study procedures (chapter outlines or underlining, lesson notes or flash cards, lists of questions and answers, using metaphors and analogies) to creating computer-based networks of ideas in which concepts introduced in the past weeks and in the present are interrelated. The interviews conducted by the first author enabled us to detect specific changes that occurred in students' thinking as a consequence, as well as the difficulties they encountered in using the new learning procedure.

When students were introduced to SemNet, they initially found it difficult to think in terms of concepts and the specific connections between them. They were not in the habit of transforming and extending their ideas, nor did they often clarify and verbalize new concepts or the connections between them. Their prior approach generally was to write sentences down, memorize them, and be prepared to use them as responses. The SemNet method of summarizing information was initially alien to the students, quite different from their previous learning procedures. SemNet prompted the students to become much more analytical about each topic than they were accustomed to being.

Students also reported that the graphic representation used in SemNet (Fig. 1), a two-dimensional frame showing a single concept in the center of the screen with all its links (labeled lines) pointing to related concepts (in 'bubbles'), was incongruent with their familiar linear way of taking notes or summarizing. Thus, students had to adjust to a different way of thinking and analysis as well as to a different method of documentation. The conflict between their old habits of studying and the new SemNet construction emerged very clearly in the case of J, even though she appeared to appreciate the consolidated skeletal framework provided by SemNet:

> [In] SemNet ,you have that one key concept, and from there you can just start branching off with a lot of information; and then on a word processor, you just have a word here and then you type the definition or examples. On SemNet you see the picture, you see the circle and then inside the key word concept and then you see all these branches of things that relate to the central concept. (J,4)

But on the other hand she transformed the notes written on the board in a SemNet pattern to "conventional" notes.

> Dr. F., she writes on the board and sometimes she writes like a net and stuff but then I don't copy the net, I make it into an outline form kind ... (J, 5)

When this contradiction was brought to J's attention, different reasons (such as the size of paper) were provided for the discrepancy between the value of SemNet that she expressed and her unwillingness to use this format in her notes. Her final conclusion was:

> I guess it seems better for me to do the other way. (J, 6)

The adjustment to the two dimensional method of documentation in SemNet was difficult:

> Because when I write in with SemNet bubbles umm, its, I tried to do that and I couldn't do it because, umm, all those bubbles had so much information I lost concentration. I couldn't concentrate on, you know, one particular thing because my eye would wander to another bubble or something like that. When I wrote,

> umm, statements, ahh, line by line, my mind and my eyes are concentrating on that one line. (J, 4)

However, J found the SemNet representation useful for obtaining an overview, either introductory or summative.

> ... but to get the whole picture of something I'd refer to the bubbles if I wanted to know the whole thing, you know everything about the digestive system, like if I was just, a you know, a student that didn't have biology or anything and wanted to know the digestive system I would go to the bubbles. (J, 4)

SemNet in its entirety was generally not adopted by students for use in other courses, perhaps in part because few students had access to computers outside of class. However, students seemed to grasp some of the intrinsic features of SemNet and incorporate them into their more familiar learning procedures, as suggested in the interview excerpts below.

Recognition of the Importance of Main Concepts

Students came to understand that in a body of knowledge, not all concepts are created equal:

> I guess SemNet showed me to try to get the main concepts ... so I do more heading and more outlining in my other lectures because then I could see what the main ideas are. (L,4)

Centrality of Connections in Learning

Concepts and the salient connections among them constitute a large part of academic knowledge. They are actually inseparable as concepts are included in each instance of a connection (i.e., each proposition) and relations (connections) are used in defining or characterizing concepts. In school, however, students tend to get the notion that concepts are more valued than connections:

> In school they stress the concept more than the connection. They'll state the connection, you know, but they won't stress the connection usually ... that is my experience in school. (M, 6)

Discrimination in Taking Notes

Students felt that SemNet influenced their way of organizing knowledge. They seem to be more critical and selective in taking notes during classes.

> I changed the way that I took notes in other classes 'cause it was less writing and it was organized and you saw the importance of the lecture instead of just writing word for word what the teacher is saying because you can write word for word and not understand what he or she said to you. (KoM, 3)

SemNet users were more concise in their discourse as well as in their note-taking, as reflected in the ratio of their words to sentences (to be discussed more below).

Holistic Picture

Students appreciated the value of SemNet for integrating ideas. This was reflected, in part, by the metaphors provided by students for SemNet, which are suggestive of the notions of connectiveness and a fuller picture. The metaphors, which included such ideas as puzzle, train, pizza and sponge, were explained on the basis of putting things together.

> Before the pieces are separate ... and you have to attach it to another piece. (M,11)

> The branch out would be the cars connected to each other. (Ja, 8)

> It soaks up what you want it to soak up.(Ja, 7)

Finally, information on changes in learning habits was also obtained with Schmeck's Inventory of Learning Processes. At the beginning of the course, the two Groups showed no significant difference on three of the four scales of the Inventory. On the fourth measure, Methodical Study, the significance was marginal ($p < 0.053$). These results (Table 1) suggest that the incoming students were quite similar in the way they process knowledge in the classroom.

Table 1. Pretest and posttest differences on Schmeck's Inventory of Learning Processes for SemNet and Comparison Groups

Group	Deep Processing		Elaborative Processing		Methodical Study		Fact Retention	
	Pre	Post	Pre	Post	Pre	Post	Pre	Post
SemNet (32)	9.93	11.68*	8.84	9.40	11.00	10.34	4.21	4.81*
Comparison (30)	10.60	11.06	8.56	8.93	9.23	9.66	4.16	4.73

* $p < 0.01$

On the posttest, the SemNet Group made significant gains on the Deep Processing scale whereas the Comparison Group did not. An increase in deep processing presumably reflects a shift toward more analytical thinking. This result is in congruence with the shifts in learning processes described above as well as with observations using Schmeck's and other means from previous semesters.

Subsequent studies by Dr. Phoebe Roeder with Schmeck's Inventory showed that many classes of the 412 series, including physics and chemistry which are taught in a constructivist fashion but without SemNet, tend to show significant gains on the deep processing and/or elaborative processing scales. However, results are not consistent from semester to semester as they have been for SemNet sections, raising questions about what this instrument actually measures.

Knowledge of Biology

The use of SemNet appeared to affect the ways in which students organized their knowledge. Towards the end of the course, the students who were interviewed expressed the feeling that their learning of biology benefited from using SemNet. They gradually realized that net construction is a process which requires them to reach into their memories and to connect new information with their prior knowledge. Further, as these ideas (that otherwise would have often remained unconnected) are linked together, the students begin to think about how to organize the knowledge cluster.

> I'm having trouble getting everything to relate, but yet I'm constantly trying to figure out how to get them to relate so I know , sooner or later, things will come to me and its a whole new way of learning, it's not the drill and practice, and it's really gaining a good knowledge and gaining a better understanding. (MS, 6)

Students also expressed their feeling that SemNet made things more concrete for them. As the concepts still remain abstract in principle, students' feelings of concreteness are a bit puzzling to us. Perhaps they are explained by the 'concreteness' defined by Wilensky [22]:

> Concreteness is not a property of a subject but rather a property of a person's relationship to an object. (It is a) property which measures the degree of our relatedness to the object, the richness of our representations, interactions and connections with the object.

It seems likely that students were using 'concrete' in the sense described above or in the closely related sense of 'familiarity with', which makes concepts more meaningful and malleable.

Metacognition

Metacognition may be defined as being aware of and exerting conscious control over one's thinking processes [21]. The question being asked in this study is, to what extent did students using SemNet acquire metacognitive skills?

One way to assess whether or how students think about their study and learning methods is to ask them to make recommendations about how to improve learning to a friend. When students were asked to make recommendations to improve the learning habits of two friends, one who was a good student and the other who was a poor student, they seemed to draw upon a common repertoire of learning strategies. At both the beginning and end of the course, most advice was of a general nature, referring to establishing regular study habits at scheduled times and places and seeking help from a tutor or peers. Other strategies included generating overviews, summaries, chapter outlines, questions and answers, or flash cards. Recopying notes, re-reading text and 'cramming' were also recommended. Metacognition would have been demonstrated if students had given advice such as, "*Think about how you are thinking; analyze your mental habits; or use concept mapping to reflect upon your knowledge*." However, such advice was very rare, as in:

> I would examine her study habits. (MB)

About one-third of the SemNet students (9/32) suggested using SemNet as a learning tool, however. This advice was sometimes followed by the reservation that it takes a lot of time. Given that this group of students tended to be computer illiterate to a large extent and few owned their own computers, it was perhaps not surprising that such advice was given hesitantly.

In addition, a significant change was detected in the importance assigned by SemNet students to the mental procedures of integration. Verbs such as *integrate*, *relate*, or *make associations* were doubled for the Group using SemNet (15 on the pretest to 34 on the posttest) whereas for the Comparison Group such references remained almost constant (15 on the pretest to 13 on the posttest).

In their interviews, students reported changes in their approaches to studying for other courses, including such shifts as focusing on main ideas. These reports reflect some degree of awareness on the part of the students about their thinking processes and how they were altering them. It seems likely that prompts provided by the interviewer helped students to articulate changes in their thinking habits, whereas the written question was not sufficiently stimulating to draw students out.

It has been suggested previously [7] that there may be several levels of metacognition and that a new vocabulary is needed to describe those levels accurately. For example,

(1) an individual's mode of thinking may first shift unconsciously,
(2) the individual may become vaguely aware of the shift in thought patterns,
(3) the individual may gradually get to the point of being able to describe the changes in thinking in response to external prompts,
(4) further introspection may be required before the individual is able or likely to describe the changes spontaneously, without specific help.

Thinking processes, like feelings, are difficult to describe. We lack a detailed language about such processes, and most students are not in the habit of thinking or talking about how they are learning.

Retention and Retrieval

As students are constructing a knowledge net, they are continuously engaged in clarifying concepts to themselves and verbalizing the connections among them. They create interwoven networks of ideas that gradually become more extensive and more elaborated bodies of knowledge. If we assume that the organization of ideas in students' minds is both reflected by and influenced by the explicit knowledge organizations they create in the computer, then, according to [1], these students should be better able to retrieve and apply their knowledge than students who did not use SemNet. Constructing propositional networks serves as a better study mode for retention and retrieval of information than traditional study practices [1].

The impact of using SemNet on a student's retention and retrieval was obtained by comparing the performance of the two Groups on the exercise described in Methods. The hypothesis was that knowledge acquired by the SemNet Group while using SemNet would be more stable and more accessible than knowledge acquired by the Comparison Group using regular study processes.

As described above, the first question assessed students' knowledge after reading a condensed paper, "Frozen and Alive", and studying it with regular pen and paper-based study procedures. This article describes physiological adaptations of animals, particularly frogs, to extreme cold so that they are able to survive freezing. Students were given as much time as they wanted to complete this task. The original article went beyond the biochemical knowledge of these students, but the condensed version was simplified somewhat and seemed to be readily comprehensible for most of the students in these Groups. When they were ready, each student was asked to summarize the article from memory.

The two Groups performed in similar ways on the "Frozen and Alive" question (Question 1, Table 2). The average number of propositions and words and the mean grades on the quality of the responses did not differ significantly. Since both Groups gave similar responses when they studied the condensed scientific article under the same conditions, it appears that the two Groups are about equally matched with respect to their abilities to learn by traditional methods and their abilities to write, and that their writing styles are similar as evaluated quantitatively.

On the other hand, the Groups differed significantly in their mean performance on the second question in which they were asked to describe the human digestive system. Both classes had studied the digestive system and had engaged in similar (but not identical) hands-on activities, including such things as examining organs from the bovine digestive system and burning foodstuffs of various types (bread, oil, etc.) to assess stored energy levels. However, the SemNet students also represented their knowledge about the digestive system in SemNet networks whereas the Comparison Group did not.

The SemNet Group wrote about twice as many sentences and used almost twice as many words as the Comparison Group. These richer descriptions of the digestive system generated by the SemNet Group (using SemNet) suggests that either their knowledge is more extensive or they are able to retrieve it more

Table 2. Mean scores per student for retention and retrieval questions

Variable	SemNet (32)	Comparison (30)
Q1. Frozen & Alive		
Number of sentences	8.22	7.18
Number of words	68.54	61.78
Ratio, words : sentences	8.11	8.59
Grade	3.78	3.63
Q2. Digestive System		
Number of sentences	10.32*	5.28
Number of words	79.06*	49.12
Ratio, words : sentences	7.94*	9.70

* $p < 0.01$

efficiently, or both. It is interesting to note that the SemNet Group also used slightly shorter propositions or sentences to describe their knowledge. The SemNet Group outperformed the Comparison Group in terms of the quality of content as measured by the number of salient concepts mentioned or the number of correct connections among them.

Question 3 was designed to assess transfer of ideas from the recently-read "Frozen and Alive" article. The transfer question was too difficult for these students and there was no evidence of transfer in either Group.

Summary

Results

It seems clear that use of SemNet did influence student learning processes. Engaging students in net construction encouraged the integration and interconnection of pieces of knowledge into more holistic knowledge structures. SemNet students learned to recognize the importance of concepts, of relations and of the process of relating. The evidence suggests that SemNet students were engaged in deeper processing than students not using SemNet, and that as a consequence they generated more comprehensive knowledge structures that they were subsequently able to retrieve from memory.

SemNet students appeared to develop more productive learning strategies that were manifested not only in the specific course in which they were using SemNet, but which also extended to their other courses. In particular they learned to look for main ideas, to clarify ideas and make them explicit, and to look for relations between ideas. The shift to more meaningful modes of learning is a difficult and prolonged process, especially among college seniors who have well-established and successful patterns of rote learning. It seems likely that if SemNet and other knowledge construction tools were used extensively at much earlier stages in students lives, meaningful learning habits might become well established and the many cognitive skills involved in learning for understanding could be taught explicitly.

Advanced Knowledge Acquisition in Ill-Structured Domains

According to cognitive flexibility theory [16], advanced knowledge acquisition is the stage in learning in which the learner must:

> *attain a deeper understanding of content material, reason with it, and apply it flexibly in diverse contexts.* [16, p. 375, italics in original)

This is the stage at which our students have arrived. If they are to successfully adapt biology lessons to various grade levels and classroom contexts, they must have a deep understanding of the concepts they wish to teach and be able to reason with them. For this reason, a primary goal of our efforts is to help students construct runnable mental models of biological processes. A coherent dynamic model can yield answers to many questions, simply by 'running' the model in one's mind.

Spiro et al. [16] identify seven conditions which facilitate advanced knowledge acquisition in ill-structured domains. We review these conditions below and attempt to evaluate the extent to which this biology course for prospective elementary school teachers meet these conditions.

Avoidance of Oversimplification and Over-regularization. When knowledge is organized functionally within a semantic network, a great deal of complexity can be captured. In this course, perhaps the most complex runnable model we create is the circulatory system, through which we trace the course of blood using the relation *receives blood from / passes blood to*. This involves following blood through chambers of the heart, to the lungs and back, through other chambers of the heart, and around the body. We take only major routes through major blood vessels and visit only major organs, but even in this simplified form, it is quite a journey when taken step by step. Creating a model like this essentially requires students to attend to complex details and to become aware of many irregularities and asymmetries.

Multiple Representations. When students organize ideas in SemNet they have finger-tip access to more than a dozen different representations of that information. Perhaps one of the most useful features is to be able to extract individual hierarchies and temporal flows from a more complex network. Another is to be able to include pictures and sketches at various points in a net. Yet another is to examine all the elements in the net via the About Net dialog. We also ask students to print key elements from their nets and map them onto large poster boards to create yet another view. We don't ask students to make systematic efforts to use integrated multiple analogies for complex concepts as recommended by Spiro, but we may soon add this to our strategies. Spiro emphasizes the importance of revisiting, reflecting upon, and rearranging ideas during the learning process, which is precisely what occurs as students create SemNet representations. Revisiting and reconsidering are significantly increased when students work collaboratively, which they have done in semesters following the studies described here.

Centrality of Cases. Specific cases or instantiations are central to learning in this course. Laboratory experiments and demonstration tasks provide the starting point for each topic. Often a series of graded tasks is employed to help students add increasing layers of complexity to their mental models, as with the diffusion and osmosis series [3]. Persuading students to incorporate references to their specific laboratory observations in their semantic networks initially proved to be difficult, however. This situation improved considerably when we designed a set of relations that students could use for linking observations and theory, derived from Gowin's Vee.

Conceptual Knowledge as Knowledge-in-Use. As noted above, knowledge representation seems to be most successful when it is organized from a dynamic, functional perspective. Designing assignments for students to lead them in this direction is not initially easy. We feel we have made significant progress with the representation of human body systems and ecosystems. Our students' semantic networks about molecular and cellular biology are still more inert than we would like, however.

Schema Assembly (from Rigidity to Flexibility). Spiro et al. [16] contend that fixed knowledge structures are not very useful in ill-structured domains because they don't fit all cases. Learners, in their view, must be able to mobilize potential knowledge and then select necessary components to fit a given situation. We attempt to build this skill by including many transfer questions on weekly quizzes. Students can tolerate very hard questions (which is typically what novel transfer questions turn out to be), so long as the questions are presented and subsequently discussed in a supportive environment and the grade penalty isn't too great.

Non-compartmentalization of Concepts and Cases (Interconnectedness). Interconnectedness, of course, is what a semantic network is all about. We have found, for example, that the *longest* distance between any two concepts in a 2300-concept biology network is eleven nodes (that is, eleven concepts). In fact, the interconnectedness sometimes swamps students initially. To manage this problem, we often encourage students to create a compartmentalized network first, as for each independent body system, and then to merge their completed nets and focus on polishing the interconnections and adding interactions between systems.

Active Participation, Tutorial Guidance, and Adjunct Support. As Spiro [16] says, you cannot simply pass advanced knowledge in an ill-structured domain to a learner. They have to construct it themselves, investing considerable effort in the process. Our class is oriented to provide incentives and supports for doing just this.

In sum, the format of this capstone biology class appears, in our judgment, to provide the kind of support needed for students to begin to make the shift to 'owning' knowledge about an ill-structured domain such as biology. Such ownership can foster students' transitions into classrooms where they are responsible for teaching biology.

Acknowledgments

The research reported here has been supported by the College of Sciences at San Diego State University, Ben Gurion University, the San Diego State University Foundation and by grants from the National Science Foundation (TPE-9053803, USE-8951817, and TPE-8950315), the Apple Classroom of Tomorrow Program, and the State of California Lottery Funds. Some of the data described here has also been reported in Gorodetsky, M., Fisher, K.M., Wyman, B. (1994), Generating Connections and Learning with SemNet, a Tool for Constructing Knowledge Networks, *Journal of Science Education and Technology,* Vol. 3, No. 3, pp. 137–144.

References

1. Anderson, John R. (1990). Cognitive psychology and its implications. Third edition, New York: W.H. Freeman and Co.
2. Feher, E. (1983). A model for preservice science training of elementary school teachers. Paper presented at the National Science Teachers Association (NSTA), San Diego, CA.
3. Christianson, R. & Fisher, K.M. (1995). In preparation.
4. Fisher, K.M., Faletti, J., Patterson. H.A., Thornton, R., Lipson, J., & Spring, C. (1990a). Computer-based concept mapping: SemNet software – a tool for describing knowledge networks. Journal of College Science Teaching, 19(6), 347–352.
5. Fisher, K.M. (1990b). Semantic networking: The new kid on the block. Journal of Research in Science Teaching, 27, 1001–1018.
6. Gunstone, Richard F. 1992. Constructivism and metacognition: Theoretical issues and classroom studies. In Research in Physics Learning: Theoretical Issues and Empirical Studies, Proceedings of an International Workshop, edited by R. Duit, F. Goldberg and H. Niedderer, IPN, Kiel, Germany, pp. 129–140.
7. Hewson, P., (Chair) (1993). Symposium on metacognition and conceptual change. Annual Meeting of the American Educational Research Association, Atlanta, GA, April. [The notion of levels of metacognition emerged in the discussion following the presentations.]
8. Johnson, D.W., Johnson, R.T., & Smith, K.A. (1991). Cooperation in the classroom. Edina, MN: Interaction Book Co. [Heller]
9. Novak, J & Gowin, D.B. (1984). Learning how to learn. Cambridge, UK: Cambridge University Press.

10. Osborne, R. J. & Wittrock, M. C. (1983). Learning science: A generative process. Science Education, 67, 489–508.
11. Pope, M. (1982). Personal construction of formal knowledge. Interchange on Educational Policy, 13, 4.
12. Pope, M. & Gilbert, J. (1983). Conceptual understanding and science learning: an interpretation of research within a sources-of-knowledge framework. Science Education, 70(5), 583–604.
13. Salomon, G., Globerson, T. & Guterman, E. (1989). The computer as a zone of proximal development: Internalizing reading – related metacognitions from a reading partner,.Journal of Educational Psychology, 81, 620–627.
14. Schmeck, R.R., Ribich, F. & Ramanaiah, N. (1977). Development of a self-report inventory for assessing individual differences in learning processes. Applied Psychological Measurement, 1, 413–431.
15. Schmeck, R.R. & Ribich, F.D. (1978). Construct validation of the inventory of learning processes. Applied Psychological Measurement, 2(4), 551–562.
16. Spiro, R.J., Coulson, R.L., Feltovich, P.J. & Anderson, D.K. (1988). Cognitive flexibility theory: Advanced knowledge acquisition in ill-structured domains. In Tenth Annual Conference of the Cognitive Science Society, pp. 375–383. Hillsdale, NJ: Erlbaum.
17. Storey, K.B. & Storey, J.M. (1990). Frozen and Alive. Scientific American, December, 92–97.
18. van Dijk, T. & Kintsch, W. (1983). Strategies of discourse comprehension. New York: Academic Press.
19. Kintsch, W. & van Dijk, T.A. (1978). Toward a model of text comprehension and production. Psychological Review, 85, 363–394.
20. White, R.T. (1988). Science Learning. Oxford: Basil Blackwell.
21. White, R.T. (1993). Paper presented at the Annual Meeting of the American Educational Research Association, in a Symposium on Metacognition, Atlanta, GA, April 12–18.
22. Wilensky, U. (1991). Abstract meditations on the concrete. In: Constructionism, edited by I. Harel & S. Papert. NJ: Ablex, pp. 193–203.

A Folding Model of Concept Genesis and Its Application to Teaching Biology

Gerhard Schaefer

Faculty of Education, University of Hamburg
Von-Melle-Park 8, D-20146 Hamburg, Germany

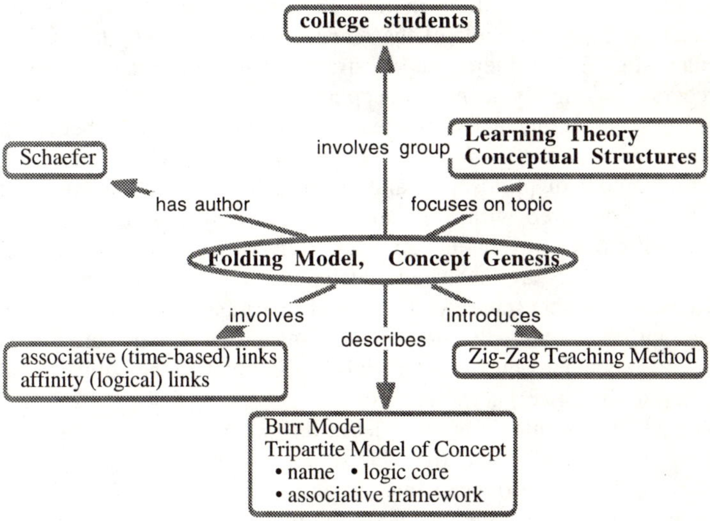

Abstract. A concept is understood here as a tripartite structure in the human mind consisting of a *name*, a standardized meaning (the *logic core*) and the *associative framework* surrounding the core. This concept of a concept is based on the assumption of a linear storage of everyday events in our memory by means of associative links (coincidence in time; Ebbinghaus), and subsequent secondary folding of the 'chain' by means of logical relations between similar events. Consequent application of this model to teaching leads to 'zigzag learning', which will be explained here as a successful method of concept formation in biology courses.

Keywords. Associative framework, association chain, free association test, folding model, logic core, protein, burr model, name, tripartite concept structure, zigzag learning, semantic network, principles of life

Introduction

Literature on concept formation published in the past 15 years (see, for instance, various issues of the *International Journal of Science Education*), as well as classical epistemology in philosophy and science, are mainly basing upon a bipartite model of a concept. A concept is seen as a twofold structure in our mind consisting of a *name* (term, designation) as the 'carrier' of a meaning, and the meaning itself. The latter can be defined as a "set of invariant attributes characterizing a special class of things" [13] and can be expressed by a definition.

This 'scientific structure of a concept has been proven to be insufficient in education [14, 15, etc.] because it is mainly *object-oriented*, and does not sufficiently reflect the subjective side of the *learner* which has to be known for instructional planning. The problem was illustrated in recent debates between concept researchers speaking about 'mis-conceptions' of students [10] and others speaking about 'pre-conceptions [2], both meaning just the same. Whether 'mis' or 'pre' depends largely on the *model* of a concept the researcher has in mind. In the old twofold model of philosophy and science, the prefix 'mis' may be appropriate because compared with a scientific definition the preinstructional ideas of students often are wrong indeed. However, with respect to the life history of a student, concepts are always in development and thus 'pre-concepts' of a later (and normally better) stage. At the time of investigation, they may even be 'appropriate' with respect to the present circumstances of the learner and his/her experience. This means they can be 'right' (according to the subject) and 'wrong' (according to the object) at the same time.

The situation leads to assuming a new model of a *tripartite* concept consisting of the *name*, the *logic core* and, in addition, a husk of associations, the *associative framework* (Fig. 1). It was introduced into literature as the 'burr model' [13] symbolizing the name by a stalk (the 'suspensor'), the logic core by the seeds in the center of the fruit (the 'vital information'), and the associative framework by the husk of hairlets and hooklets which attach the fruit to carriers and allow transport to different places of use.

In this paper some theoretical arguments will be given to justify the threefold concept model from a specific notion of concept genesis, and also first empirical results will be reported supporting this theoretical view.

Theoretical Part: The 'Folding Model'

The theoretical basis of the burr model has to be seen in (a) early investigations on the process of learning made by Ebbinghaus [6], (b) experiences gained from association tests in clinical interviews made by psychoanalysts (Freud, Jung, Adler), (c) epistemological studies carried out by Deese [4] highlighting the role of associations in language and thought, (d) reflections of De Bono [3] on the promotion of creativity by 'lateral thinking', and finally (e) a number of concept

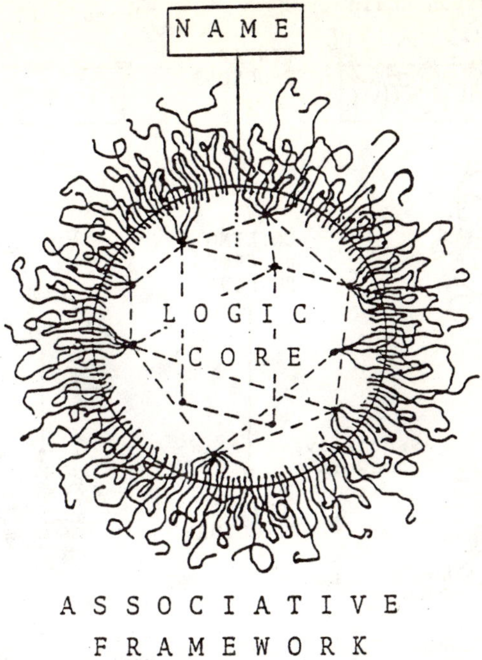

Figure 1. Burr model of a concept consisting of three parts: the *name* (designation), the *logic core* (object-specific meaning) and the *associative framework* (subject-specific connotations) [13]

studies undertaken by the author in 20 years' research [e.g., 5, 12, 13, 14, 15], which deal with the logic core as well as the associative framework of concepts and their influence on human behavior.

The old theory of Ebbinghaus, which never since has been falsified, can be taken as a valid theory for concept understanding. It says that everything in life that is in close proximity in time is linked in memory just by associative bridges independent of any logical relationships. Even pure 'nonsense links', like the smell of a flower and a traffic accident, are associated if they coincide in time, and the two elements are able to retrieve each other from memory later on.

The idea now is that an infinite chain of 'main events' (events in the center of awareness) enter our memory, and lateral to this chain there are numerous 'side events' (association clusters) stored likewise (Fig. 2, left side). Then, in a subsequent process, certain elements of the chain which are similar either in structure or meaning or affective quality, are additionally linked by *affinity relations* whereby the chain is folded or curled up from its primary linear structure into a secondary 'spatial' structure (Fig. 2 left, middle and below). Admittedly, this happens anyway within the spatial structure of the brain, but we do not know yet whether the model of 'folding in space' is not just a metaphor.

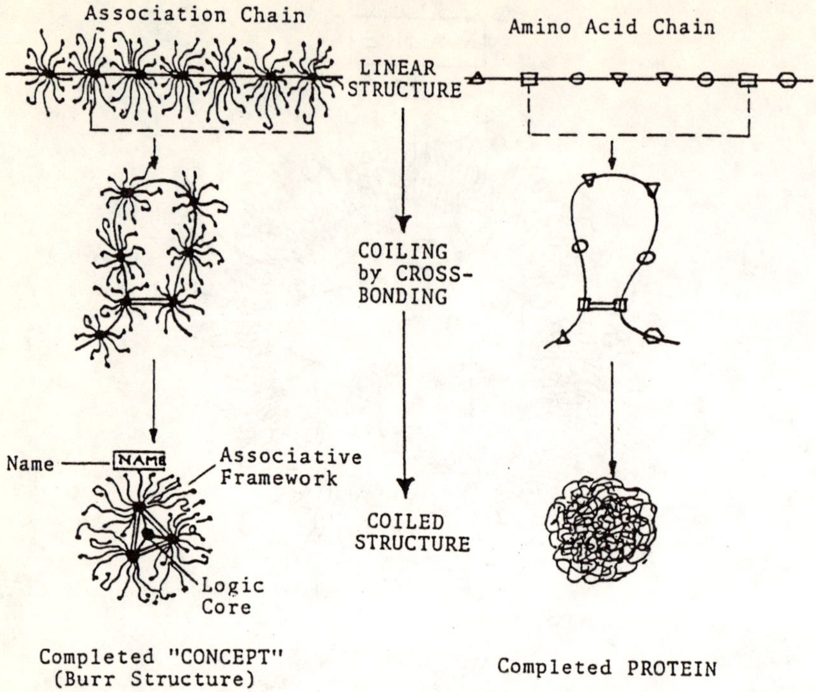

Figure 2. Folding model of concept formation in schematic survey [19]. Left: Development of a concept from a primary association chain by secondary affinity bonding. Right: Biosynthesis of protein for comparison; development of a protein molecule from a primary polypeptide chain by secondary sulfur or hydrogen bonds.

Also the postulated sequence of primary (associative) and secondary (affinity) bonding, although rather plausible, is a simplifying hypothesis so far. The possibility exists that the associative (= consecutive in time) and the affinity (= logical) links are formed almost simultaneously: Prolongation of the chain and its cross-folding may follow each other so fast that the two processes appear as parallel.

By repetition of this process, i.e., renewed storage of similar events in the chain under changing conditions (e.g., big dogs, small dogs, white dogs, black dogs, curly dogs, smooth-hair dogs – all with the same basic properties of the species *dog*), a structure of permanent 'knots' with permanent logical relations is developed, the 'logic core' of the concept. It is surrounded by numerous loops of the rest of the association chain and their attached association clusters which form the 'associative framework' of the concept.

One of the repeated associations is the 'name', the designation of the concept. Its links to the concept structure are sometimes associative (if the name was chosen incidentally and then maintained by social acceptance), sometimes logical (if the name was introduced purposely because its parts express best the logic core of the concept). Insofar the name appears somehow between the associative framework and the logic core when looking from the perspective of concept genesis. Its clear distinction from both of them, however, is possible from the perspective of function: The two functions of a concept name (see the stalk in the burr model) are (1) *codification* in memory, and (2) *communication* in the language group, whereas the function of the associative framework is the *adaptability*, and applicability, of the concept to different situations of life, and the function of the logic core the appropriate depiction of a type of reality.

This model of concept formation was first mentioned in 1985 [5], but explicitly explained in 1990 [19]. It should be mentioned here that the designation 'concept' for this psychological structure is an excellent choice. The Latin word *con-cipere* (participle perfect: *con-ceptum*) means 'taking together, grasping together', and this is actually what happens in our memory: The endless thread of associations is coiled up to various knots, burr units, 'concepts', and these again are interlinked by logical and associational relationships and form a highly complex structure in our mind which is composed of 'epistemic' and 'sporadic' parts (see Fig. 3).

The whole process can be compared with, and symbolized by, the biogenesis of protein (Fig. 2, right side). Here we also find a one-dimensional, linear structure which is folded up to secondary and tertiary forms, new compact units which are physiologically active. Theories of memory based on the assumption that protein synthesis plays a role in the storage of information lead us to the idea that there might be more than just an analogy between concept formation and protein synthesis.

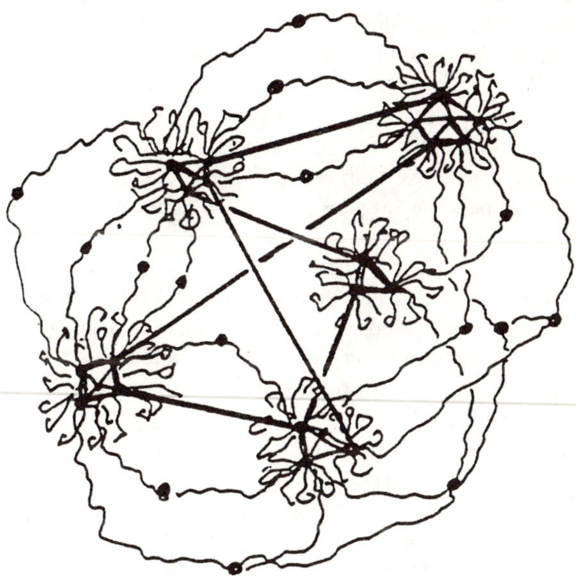

Figure 3. Network of concepts in the human memory built up by logical and associational links (mixed epistemic/sporadic structure).

An interesting approach to the understanding and modeling of concept formation was recently published by Giordan [9]. His 'allosteric model', however, apparently under-estimates the role of association chains in concept development and – a typical trend in modern cognitive psychology – gives logical relations between concept elements priority to associative relations. Here we find again the usual way of concept understanding: What is secondary according to the folding model, is regarded as primary, and vice-versa.

If we check the immense number of publications on concept research in the last 15 years, it is striking to see that most of them deal only with the logical, 'scientific' part of concept relations and very few with associations. Only recently, in the last 5 years, an increasing number of investigations using methods like concept-mapping or semantic networking (for the methods, see Fisher [8]) pay attention more to the association side of concept structures without, however, using this opportunity to elaborate a fruitful theory of concept formation.

A promising step into this direction, to the authors opinion, has been made by Fisher [7] who included associations as 'unnamed links between concepts' into the structure of knowledge which then, after the meaning of the link has become clear, are transformed into 'named relations'. In this approach, a 'sporadic' knowledge structure, tied together by associations, can principally be converted to an 'epistemic' structure built up on logical relations. In terms of the burr model this means: the logic core expands into the associative framework whenever associations are becoming conscious and explicable as logical relations. This sounds reasonable, but contradicts in some way the essence of our folding model: according to the latter the elements of the logic core result from *repeated similar* associations, whereas in Fisher's approach also *single* associations can become 'logical'. If we accept this (and there is no reason not to do so), we have to postulate two different kinds of logical relations: those belonging to the 'logic core' and thus being 'essential' for the understanding of the type of reality which the core stands for, and those belonging to the concept frame with its sporadic elements. They may become 'logical', but still remain irrelevant for the concept meaning.

The theory behind Fig. 2 is that the primary step in concept formation is the *associative linkage* of events, and that the logical connection between them is a secondary step only, performed by an additional operation in the brain. Both steps may occur (and certainly very often do so) unconsciously so that concept formation, in principle, is a subliminal process. This explains why we find concepts (non-verbal and verbal) already in early childhood, before any intentional education by parents or teachers takes place, and why the existence and function of non-verbal concepts has been proved also in higher animals [11].

If thus a concept, as is the case in philosophy and in the disciplines of science, is understood as a bipartite structure only consisting of name and (object-specific) 'meaning' while the associations around name and meaning are neglected because they do not 'matter' in this context, the genuine nature of a concept is just turned upside down. Admittedly, the associations do not belong to the 'matter', i.e., to the abstract type of repeated events which we call the 'logic core'. However, they belong to the person who spins the thread (the 'Ebbinghaus chain'). Since this thread, according to our theory, is the raw material of concept formation, they definitely have a strong influence on the way concepts are formed in our mind.

We conclude: If a scientist is interested only in the logic core of concepts, he/she will be satisfied with the bipartite model and work with it perfectly. If somebody, however, is interested in the *genesis* of concepts, and their function in *daily life*, he/she does well to use the tripartite model and to include associations in concepts. It is obvious that teachers and teacher trainers belong to the latter group, and also those persons who want to forecast other people's reactions in specific situations of life (politicians, economists, advertisers, etc.) because the human being, in the majority of cases, apparently behaves associatively rather than logically.

Furthermore, the bipartite concepts of science are often regarded by students as being 'naked', 'empty', 'dead', 'skeletons', far away from their personal life and their everyday problems. It is the richness and flavor of associations around the skeleton that form the 'flesh' and make the concept attractive to everybody, as could be proved with concepts like energy, equilibrium, order, health, life, and others.

Empirical Part: 'ZigZag' as a New Method of Teaching Concepts

For several years now the author has based his teacher training seminars at the University of Hamburg on 'universal principles of life' [16, 17, 18], in order to teach his students a 'science *of* organic life, *for* human life' which should be the main focus of syllabuses in the coming years [18]. The principles are visually demonstrated to the students partly by slides, and partly by living objects shown during excursions into the environment, the Botanical Garden or greenhouses. One week after each excursion, at the beginning of the next seminar, free associations are collected from the students given about ten stimulus words which are referring to either 'central phenomena' (CP = phenomena which where in the center of interest and instruction during the last excursion), or to 'marginal phenomena' (MP = phenomena at the margin of interest and instruction which where mentioned incidentally only). The aims are, first, to find out relations between CP, MP, free associations and the principles of life (*research aim*), and second, to train students according to the 'folding model' of concept formation (*teaching aim*, see Fig. 4).

Two hypotheses guide these seminars:

(1) In spite of an unsystematic, sporadic presentation of phenomena, the students build up an orderly network of associative and logical links between events in their minds. This results in a reasonable structure of concepts and a comprehensive understanding of life in general, and the specific environment in particular.

(2) The concepts thus developed are expected to contain a precise logic core, sufficient for scientific reasoning, as well as a rich and multidimensional

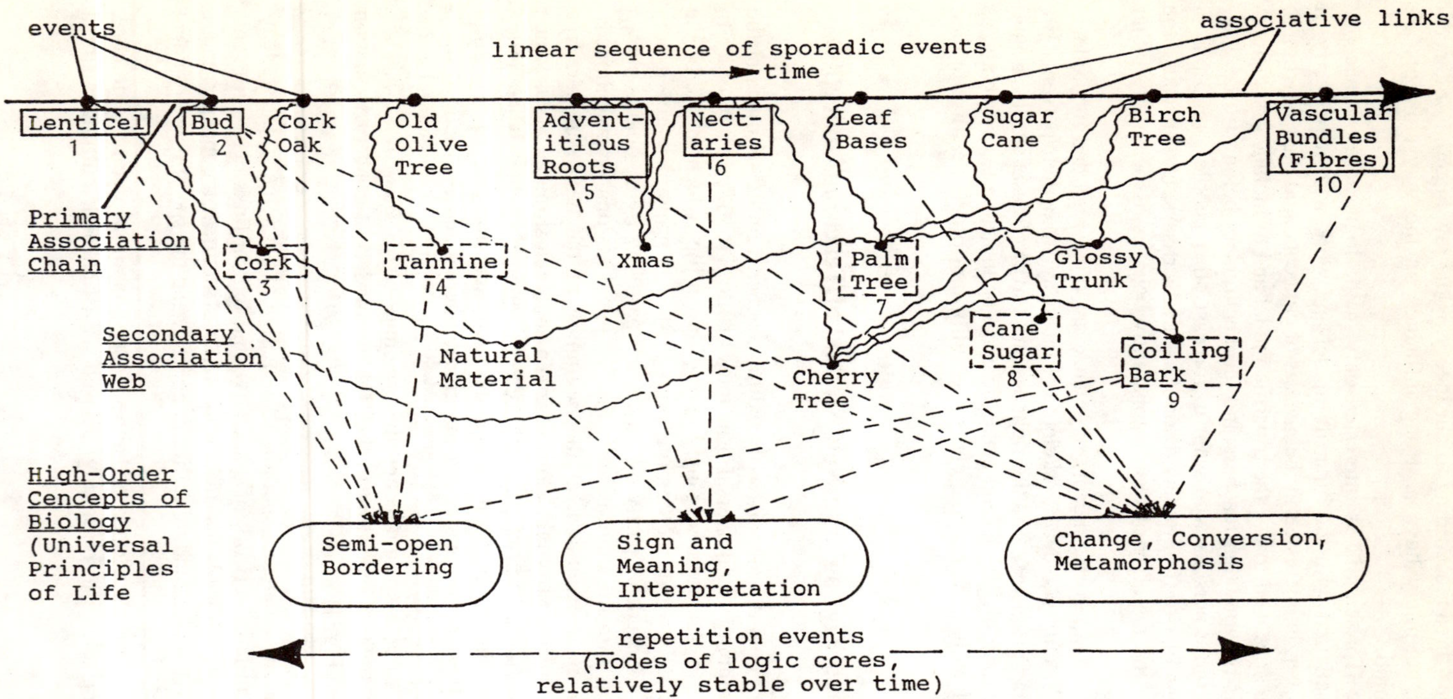

Figure 4. Process of concept formation during excursion, subsequent free association testing (1 week later, see Table 1) and parallel continuous development of 'repetition nodes' (see Fig. 5). The stimulus words used in the association test are classified as Central Phenomena (CP), indicated by solid squares, and Marginal Phenomena (MP), indicated by dotted squares. Primary associations are indicated by solid lines, secondary associations by wavy lines, and affinity or logical associations by dotted lines.

Table 1. Associations given one week after an excursion through the Botanical Garden (6 weeks after introduction to the principles of life) to ten stimulus words (listed on left). Stimulus words are classified as Central Phenomena (CP) or Marginal Phenomena (MP). All stimulus words were taken from *this* excursion and not from previous excursion or the introduction to the principles of life. Response words are indicated by abbreviations across the top.

STIMULUS WORD		CP	PL	L	S	T	Ec	Bo	An	Mo	Ay	Ph	Ch	Pr	N	Geo	M	Ps	Pol	Ma	F	Es	WA	Total: Stimulus Words
Cork	MP	12	18	–	15	–	–	4	–	4	5	3	–	–	–	9	26	–	–	–	–	–	1	85
Adventitious Root	CP	–	15	–	30	–	3	10	–	24	–	6	–	–	–	3	–	–	–	–	–	–	10	101
Bud	CP	–	15	3	12	11	2	15	–	23	1	10	–	8	–	–	1	–	–	–	3	–	–	104
Tannine	MP	3	12	–	3	–	7	17	–	1	–	11	7	12	–	–	24	–	–	1	–	–	–	95
Lenticel	CP	–	10	–	18	–	–	18	–	24	6	17	–	1	–	–	–	–	–	–	–	–	2	96
Palm Tree	MP	3	7	–	10	–	4	10	3	18	10	–	–	4	9	10	6	–	–	–	–	–	3	94
Cane Sugar	MP	12	6	–	14	–	–	10	–	8	–	13	22	–	–	6	8	–	–	–	–	–	–	87
Vascular Bundle	CP	–	4	–	10	–	1	7	–	4	41	32	–	1	–	–	–	–	–	–	–	–	1	101
Coiling Bark	MP	25	2	–	25	–	22	4	2	16	–	14	–	4	2	4	–	–	–	–	–	–	–	95
Nectaries	CP	–	–	–	6	1	11	12	20	20	–	10	–	7	–	–	9	–	–	–	–	–	5	101
Total: Categories	Total CP		44	3	76	12	17	62	20	95	48	75	–	17	–	3	10	–	–	–	3	–	18	503
	Total MP		45	–	67	–	33	45	5	47	15	41	29	20	11	29	64	–	–	1	–	–	4	456
					143!			107!		138!		116!												

associative framework. This allows quick transfer to different contexts of daily life and also gives a high motivation to the students.

Percentage of Associations in Different Categories

In Table 1 the result of an excursion in the middle of the semester is represented. The principles of life were explained theoretically and illustrated by slides 6 weeks ago, so that it may be assumed at least some of them were still present enough in the memory of the students to serve as *possible* nodes in epistemic structure constructed by the students. Additional nodes would be formed by other affinity relationsof their association chain.

The numbers in the table represent percentages of associations in this category as related to the total number of associations in the sample. The order of stimulus words in the table has been chosen according to decreasing percentage of Principle of Life (PL) associations, because one of the final aims of this study was the discovery of associative networks round the *principles of life*. The original sequence during the association test was: lenticel – bud – cork – tannin – adventitious root – nectaries – palm tree – cane sugar – coiling bark – vascular bundle.

As previous investigations showed, in zigzag learning, with its unsystematic structure of subject matter, the sequence of stimulus words applied in the association test has no visible influence on the result, that is, on the distribution of associations over the categories.

Categories CP and LP were fixed before the evaluation according to the aim of the investigation, all other categories were empirically found or theoretically introduced during the process of evaluation. Meaning of abbreviations: MP Marginal phenomenon, CP Central phenomenon, PL Principle of Life, L Life in general, S Situation during excursion, T Time of the year, Ec Ecology, Bo Botanical object, An Animal, Mo Morphology, Ay Anatomy, Ph Physiology, Ch Chemistry, Pr Property of object, N Natural phenomenon, Geo Geography, M Man (culture, society, etc.), Ps Psychological phenomenon, Pol Politics, Ma Mathematical/geometrical aspect, F Formal term, Es Esthetics, WA Word association.

As table 1 shows, quite a number of free associations refer to the principles of life although the stimulus words themselves have no apparent relation to them. The high percentages of 'cork' through 'lenticel' are mainly due to connections made with the *border principle* (shelter by cork, bud scales, tannin; opening of borders by lenticels), to a smaller part with the *convertibility principle*: bud scales as converted leaves, adventitious roots as converted shoot parts.

In the table which contains also marginal phenomena (MP) as stimulus words, the percentage of associations which refer to the corresponding *central* phenomenon (CP) is striking: on the average 11%. Coiling bark with 25% (—> 'birch tree', 'cherry tree'), cork with 12% (—> 'cork oak') and cane sugar with 12% (—> 'sugar cane') are here on top whereas palm tree and tannin, with 3% each, reveal a minimum of reference to their CP (—> 'coating by leaf bases', —> 'old olive tree'). Not only were the MP keywords visibly associated with the central phenomena, but they also stimulated a rich spectrum of further associations (15 different categories, see lowest line of the table) which is comparable to the effect of CP keywords (also 15 different categories, see lowest but one line). This pertains, too, to the Principles of Life (PL) column: 45% of all associations referring to the principles are released by MP, and 44% by CP keywords. This shows that the marginal phenomena, demonstrated or just mentioned during the excursion, contribute essentially to the building up of an association chain which later on, by folding up, yields nodes such as the 'Principles of Life'.

Table 1 further indicates that in all (biological) categories characteristic of the *central* phenomena (high CP totals; see S, Bo, Mo, Ay, Ph), the MP totals are clearly smaller. On the other hand, MP totals are dominating in Ec, Ch, Pr, N, Geo, M (!) and Ma – categories which go beyond the areas of biology. This result proves that marginal phenomena are helpful factors for the enlargement and enrichment of the associative framework even of *biological* concepts. In particular, all relations between biological objects and *Man* (category M) seem to be essentially strengthened by the inclusion of such marginal phenomena in the excursion. If we want to teach 'biology *of* organic life *for* human life' (the aim expressed at the beginning of this paper), we do well to organize the teaching process accordingly, i.e., to integrate phenomena and remarks referring to the life of humans into scientific instruction as often as possible. This knowledge is not new, and successful teachers have always followed this rule, perhaps sometimes intuitively only. However, the rule is empirically confirmed here by simple association tests made on 'zigzag learning'.

Table 1 is the result of a small group of 28 biology students. It represents a 'case study'. Statistical proof on a larger sample is, in principle, not possible because each excursion is too different from the previous one according to varying composition and size of classes, weather, season, vegetation, etc. There are never 'reproducible conditions' in zigzag learning as required for statistical analysis. And still, the results described here have been repeatedly observed through 38 subsequent case studies since 1985, with a total number of 926 students, and there is no doubt, so far, about their general validity.

Semantic Networks at the End of the Course

From the association pool collected during the semester, out of 10–12 excursions (approximately 16,000 associations!), a sample of about 100 is selected and presented, together with all stimulus words of the association tasks, on an A3 sheet in complete 'disorder' (i.e., no order familiar to the students). The students are asked to work on this 'bound association test' most elements of which are unknown to them. They draw lines between such words which they think are, or should be, associated with each other in *some* way (first step). The sheets are collected then, copied (for research purposes), and one week later handed out to the students again, with the task to signify the drawn association lines by logical relations (second step; see Fig. 5).

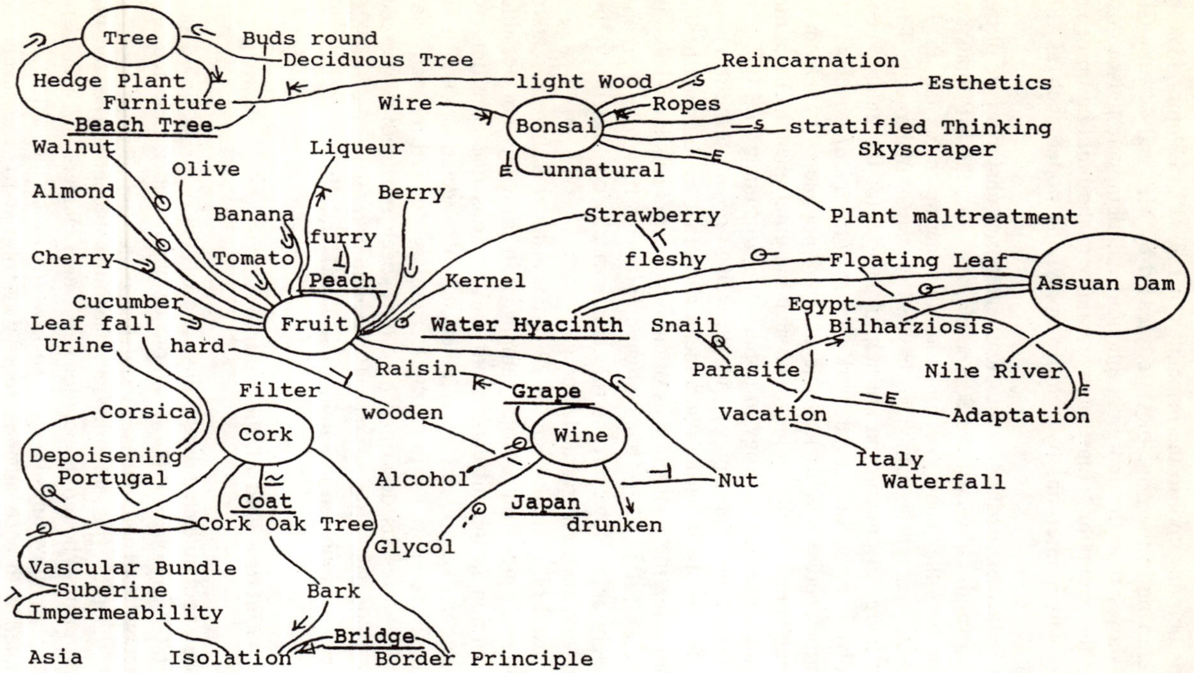

Figure 5. Semantic network produced by a student of biology from a bound association test. Free associations have been selected from the total sample over one semester (here a small section only). Connecting lines (associative links), and one week later the symbols (logical links) have been chosen by the student. Meaning of some symbols:

—͐ is a special case of; —ɵ is part of; → produces; —┤ is; —s symbolizes; —ᴇ is example of; ≃ is like; etc.

Varying the normal way of concept-mapping or semantic networking, we sometimes do not ask the students to use *words* for signifying the relations, but abstract *symbols* (see legend of Fig. 5). This may be regarded as an additional difficulty in the test procedure, hence an additional source of error and also a handicap for younger students below the age of 16. However, it could be shown in our case studies that the students, once they had practiced the symbols in a pre-exercise and were acquainted with them, handled them quite skillfully. Above all, the number of relations which can be signified in such a way, in the time given, seems much greater than by using words. The evaluation showed that the students marked (i.e., discovered) logical relations between things which had never been mentioned or defined before, during the semester. The logical relations apparently 'grew' out of the network of associations developed in the course of study without any external help from the teacher.

These results will be further pursued in order to find out more about the strategies the students apply when formulating logical relationships between associations. There is evidence that through this method we are going to get a deeper understanding of the folding model, in particular the relevance of 'affinity bonds' and their coiling effect on the association chain.

Looking at Fig. 5 again, which is only one example out of 38 case studies, and out of the final A3 sheet only a small section, we observe that the *stimulus* words used in the association test (underlined in the figure) generally are not the centers of node formation in the asociation network. On the contrary, these words have been associated, by the students, to one or two others only, whereas some other elements (quite 'unimportant' ones, with respect to the central or marginal phenomena of the excursion!) are connected with many lines to other words, thus forming a striking cluster in the semantic network. (Compare, for instance, the 'solitary' stimulus words Beach Tree, Peach, Water Hyacinth, Japan, Coat, Bridge with the strongly intertwined 'nodes' Bonsai, Assuan Dam, Wine, Fruit, Cork).

These results hint at the importance of researching further on the subject of associations in daily life and in teaching. The role these junction points of the associative network play in human actions must be examined on one side, and in the formation of logic cores on the other.

Conclusion: The Zigzag Method of Concept Learning

The teaching method described above can be summarized in seven steps:

1st step. *Unsystematic walk* through environment, laboratory, museum, etc. Objects of ad hoc interest of the learners are taken as a focus of talks and observations ('central phenomena', CP). These are commented upon by side remarks and side observations ('marginal phenomena', MP) made around the objects of focus and viewing them from different perspectives.

2nd step. One week (or one lesson) later: *Free Association Test* on about ten selected stimulus words referring to central as well as marginal phenomena of the last excursion. Test-sheets are collected.

3rd step. *Evaluation* of the free associations:
(a) Grouping in categories (see, for example, Table 1);
(b) Checking whether associations given to MPs also contain CPs, or refer to them;
(c) Checking whether associations given to CPs or MPs also contain PLs, or refer to them.

4th step. At the end of course: Selected associations from the whole course, together with stimulus words —> *Bound Association Test*. Learners draw connecting lines (*associative links*). Test-sheets are collected. Number of links, associative clusters ('nodes'), etc., are checked.

5th step. One week (or one lesson) later: Learners receive their own test-sheets back and signify the association lines with words or abstract symbols as *logical relations* —> Semantic Network.

6th step. *Evaluation* of Semantic Network. Number of symbols, correctness, etc., are checked.

7th step. *Presentation of results* to the learners. Discussion. Explanation of results by the learners themselves. Definition of newly developed concepts on the basis of all preceding teaching steps. Prospective ideas for future teaching and learning.

If many associations to the principles of life (category PL) are introduced to the students immediately before the course, more and more interconnections between *new* association chains appear during the following excursions. Finally, an increasing number of cross-relations are found to *old*, preceding stimulus words and to their corresponding association chains, thus indicating the development of a tight, multi-dimensional *web of associations*.

Within this sporadic, 'chaotic' web a secondary structure of *association clusters* originates, obviously by itself and without any directives from the teacher. These seem to develop 'logic cores' of concepts. In a time of growing importance of chaos theory, the idea may be seriously considered in educational sciences to feed the brains of learners with more associative, sporadic elements in order to provide a basis for richer concepts with greater suitability to everyday life. Zigzag learning has proved to be a useful method.

References

1. Ausubel, D.P., Novak, J.D., Hanesian, H.: Educational Psychology: A cognitive view. New York: Holt, Rinehart and Winston 1978.
2. Boschhuizen, R., Brinkmann, F.G.: Research methods on pre-instructional ideas in biology education. Proc. 13th Conf. ATEE, Barcelona 1988.

3. de Bono, E.: Lateral thinking. New York: Penguin Books 1970.
4. Deese, J.: The structure of associations in language and thought. Baltimore: John Hopkins Press 1965.
5. Deubner, W., Schaefer, G.: Wenn in die Natur noch alles in Ordnung ist. Köln: Aulis 1985.
6. Ebbinghaus, H.: Über das Gedächtnis. Untersuchungen zur experimentellen Psychologie 1885. Reprint Wiss. Buchges. 1971.
7. Fisher, K. M.: Semantic network theory for modeling conceptual change. Unpublished manuscript San Diego (personal communication only) 1992.
8. Fisher, K.M.: Semantic networking: The new kid on the block. J. Res. Sci. Teaching 27/10, 1001-1018, 1990.
9. Giordan, A.: An allosteric learning model. From the categorization of learners' conceptions to an optimal didactic environment. Document LDES, Univ. Genève 1989.
10. Novak, J.D, (Ed.): Proceedings of the Second International Seminar on Misconceptions and Educational Strategies in Science and Mathematics Education. Vol. II. Ithaca: Cornell Univ. 1987.
11. Rensch, B. Gedächtnis, Begriffsbildung und Planhandlungen bei Tieren. Berlin: Paul Parey 1973.
12. Schaefer, G.: Kybernetik und Biologie. Stuttgart: Metzler 1972.
13. Schaefer, G. Concept formation in biology: The concept 'growth'. Eur. J. Sci. Educ. (now: Int. J. Sci. Educ.) 1/1, 87-101, 1979.
14. Schaefer, G. Concept formation in school between scientific and community demands. In: Biological education for community development (P.J. Kelly, G. Schaefer, ed.), pp 127-147. London: Taylor and Francis 1980.
15. Schaefer, G., Hernandez, D. (Eds.): Health education through biology teaching. IUBS-CBE monograph. Quezon City: Univ. Philippines, ISMED 1985.
16. Schaefer, G.: Teaching biological principles – a step towards positive health education. ICSU/UNESCO conference on Science and Technology Education and Future Human Needs, Bangalore 1985.
17. Schaefer, G.: Twelve principles of life as a basis of ethics and education. In: New challenges for biological education (Schaefer, G., Younes, T., eds.). Biology International, Special Issue 11, pp 3-13. Paris: IUBS 1986.
18. Schaefer, G.: Die Entwicklung von Lehrplänen für den Biologieunterricht auf der Grundlage universeller Lebensprinzipien. MNU 43/8, 471-480, 1990.
19. Schaefer, G.: Zickzack-Lernen als Methode; oder: Kann aus Wirrwarr Ordnung entstehen? In: Erleben, Beobachten, Untersuchen. Zur Didaktik von Exkursionen (B. Oehmig, ed.), pp 110-131. Berlin: FU, Zentralinst. f. Fachdidaktiken, 1990.

Biological Interrelationships and Water

Michael J. Brody

College of Education, Montana State University, Bozeman, MT 59717, USA

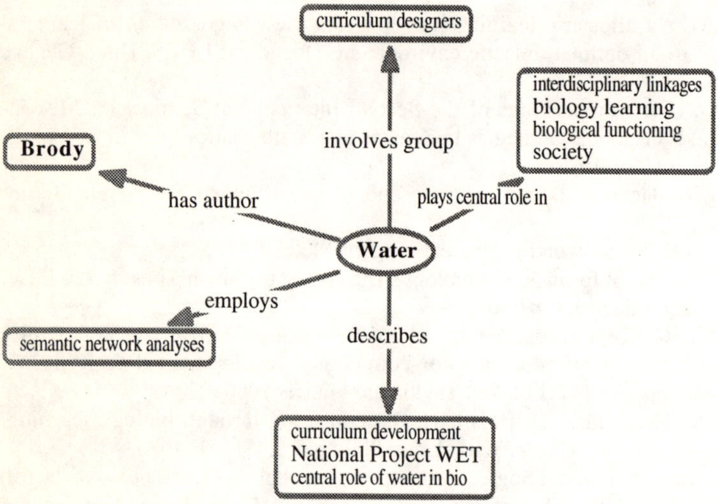

Abstract. Water is a critical concept in science and can serve as a conceptual link among various topics in biology as well as in a variety of both natural and social sciences. This paper describes analysis of the central role of water in learning and teaching about biology. The role of water in the structure of biology and ecology knowledge is analyzed through the construction of a semantic network created by the author using SemNet software. Several water-related concepts indicating significant relationships among various areas of biology and other disciplines are described. The need for continued restructuring of disciplines in order to design more effective curriculum and promote meaningful learning is discussed.

Keywords. Biology, ecology, epistemology, learning, curriculum

Introduction

Increasingly scientists, educators and the general public are concerned about the quality of science education in schools. This concern is compounded by a growing realization that the earth is a finite system and the future of people and civilization is threatened by numerous ecological crises. The integration of basic science concepts and skills with real life concerns, including the future of the planet, is critical to help maintain the biosphere and quality of life.

In order to become responsible citizens, students must learn that people are a critical part of the environment, they respond to changes in the environment and increasingly cause rapid changes to the environment. The roles that science, technology and societies play in our daily lives and their implications for future generations is an integral part of an effective educational system.

Recent international meetings focusing on ecology and the state of the world confirm a growing realization that people must be educated in a holistic fashion with a concentration on real life issues that affect life on the planet. Economic and political interests must be integrated with ecological concerns. World issues such as global warming and biodiversity are essential for citizens to understand. Educators must take into consideration that children's understandings of science are influenced by cultural beliefs. For any individual these understandings are modified by psychological and personal concerns, some of which are related to health, diet and disease.

There is a clear need for an integrated, holistic approach to science education with a focus on issues which will affect the world's population. The problem remains: how to accomplish these goals and what strategies will be successful. One approach is to reorganize the school curriculum around concepts which have the greatest potential for bridging the gaps between disciplines. Of all the possibilities available to educators at this time, it appears that no one path holds the greatest promise but rather combinations of existing and new strategies will be most valuable.

National Project WET (Water Education for Teachers) has taken on the task of developing curriculum resources and educational materials focusing on water and its use by people. The project's goal is to provide teachers with conceptually integrated materials for grades K–12.

Nature of Knowledge About Water

The concept of water (which in its totality may be called the hydrosphere) may be the most integral of all concepts related to life and the earth and thus is critical to achieving an understanding of the complexity and interrelatedness of earth systems. Without water, life would not exist as we know it. From the molecular to the global level, it is what makes the earth unique.

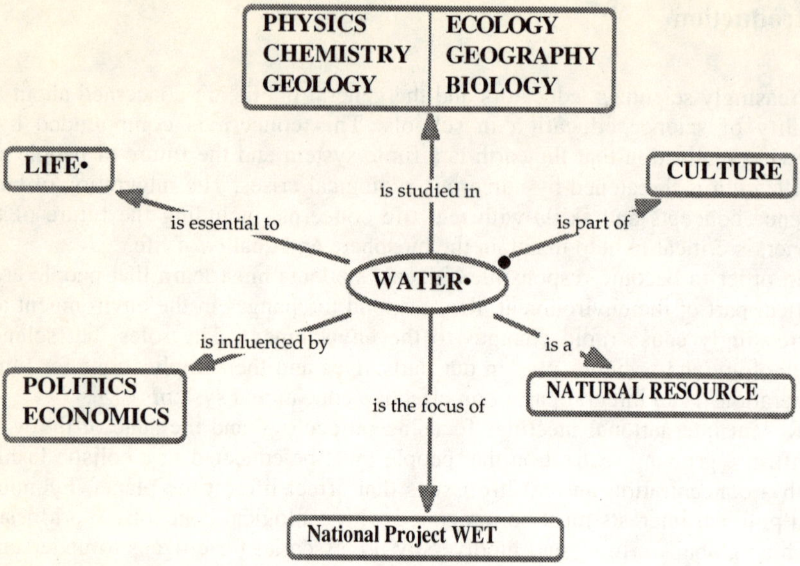

Figure 1. Opening frame of semantic network focusing on water showing interdisciplinary nature of concept.

Ancient civilizations considered water to be one of the four elements: the earth, the wind, the fire and the water. The Greek scholar Thales of Miletus in the beginning of the sixth century claimed water to be the primordial substance, that from which all life arose and to which it will eventually return. Today we know that people can live for weeks without food; however, they can only live a few days without water. The dependence of people on water is apparent at many levels, from water in the air to help keep our lungs moist to providing industrial energy as well as recreational resources.

If education is to move into an integrated and holistic mode then we must seek clear examples of concepts which can be used to bridge the so-called traditional disciplines such as biology, chemistry, social studies, mathematics and art. Water provides such an example.Water is integral to all life processes and thus central to the study of biology. It is powerful in its ability to relate all disciplines in interesting ways. In the natural sciences physics explains the structure of water, its electromagnetic configuration and molecular bonding which leads to its unique properties. Chemistry helps us understand the role of water in all life processes. Water distribution is a major component of geological sciences. Biology and the study of life is based on an understanding of water and all ecological processes are water related. The social sciences are also inextricably linked with our understanding of water. Ancient and modern civilizations grow and collapse based on availability of water for their people. Economics is based on access to water. Manufacturing and agriculture are water dependent and politics is often invoked to legislate water resource availability. Natural resource management strategies for water such as dams have a significant influence on energy, transportation, recreation and commercial interests. Water can be considered a continuous web which unites the entire earth and thus an understanding of its role can lead to a new meaning of interrelationships on earth.

Epistemological and Psychological Frameworks

The research was guided by the following principles: (a) before attempting to design instructional activities in any domain, the major concepts and organizing principles of the knowledge domain should be identified [7]; these principles should be broad and inclusive, stressing conceptual relationships and meaning rather than isolated facts, (b) the assessment of knowledge through conceptual analysis provides a comprehensive picture of concepts and conceptual relationships [10], and (c) the assessment of knowledge in a given domain provides information useful in the design of curricula and educational materials that help to introduce new and difficult concepts in ways that can facilitate meaningful linkage of those concepts to existing relevant knowledge in students' cognitive structures [1].

These epistemological issues related to the nature of knowledge about water are approached from the constructivist paradigm. Constructivism follows in the wake of the death of positivism as the major guiding belief in science and science education. What has been viewed as the objective production of knowledge is now seen as a more human endeavor subject to personal beliefs and idiosyncratic perspectives on natural phenomena. Scientific models are viewed as evolving over time, restructuring old ideas and integrating new ones. Much like the evolutionary perspective of life, the production of knowledge as seen through the constructivist view, is one of continuous gradual change modified by episodic events similar to punctuated equilibrium.

Given this rather fluid perspective of knowledge, new strategies have been developed to represent it in ways which allow the researcher, teacher and learner to modify and evolve knowledge structures. Thus concept mapping can be used to organize and represent knowledge in this way [12]. Given the complexity of even one discipline such as biology, much less the integration of several disciplines around the concept of water, it is necessary to identify concepts that are integral to understanding and to relate these concepts in meaningful ways. Concept maps have been useful in this task [6]. However the two-dimensional nature of concept maps generated on paper can be limiting. Many of the most interesting and complex relationships simply cannot be expressed in two dimensions. Problems such as these have led to the use of computer-based concept mapping techniques which allow the person constructing a knowledge network to operate in *n* dimensions [9].

Previous Work

The analysis presented here is based on the premise that learning is the comprehension and acceptance of concepts which are intelligible and rational and lead to a change in the meaning of experience for the learner [11]. Based on the constructivist paradigm, consideration is given to what people can learn as an integrated set, conceptual framework or cluster of concepts related to water rather than understanding a single concept. This reflects a complex, integrated and multidisciplinary conception of a natural phenomenon. In several studies concerning children's conceptions of the world, we have found that students seldom see the multiple connections that link science to the real world. Students are unable to explain higher order concepts such as **aquatic ecosystems**, **nutrient cycling** or **productivity** in terms other than those provided in their text [5].

Three studies related to the concept of water were selected following news media surveys of then-current ecological crises. The key issues were children's knowledge of the **international maritime boundary dispute** between Canada (CN) and the United State of America (USA) in the Gulf of Maine, **acid deposition** and **pollution**, all of which relate to the concept of **water** in interesting and complex ways.

The Gulf of Maine maritime boundary is located in a uniquely productive and valuable marine resource which is shared by the USA and CN, and which has significant impact on the social, political and economic fabric of both countries [3]. The 1984 World Court hearing concerning the maritime boundary brought together a mass of fundamental and current information on numerous science and natural resource concepts related to water. The World Court proceedings and related scientific research on the Gulf of Maine served as a framework for understanding the nature and management of marine systems. Among the concepts related to water which were in investigated in this study were: **physical** and **chemical characteristics**, **dissolved gases, nutrients, currents** and **upwelling** (Fig. 2).

A similar study of marine resources [6] conducted in the state of Oregon (USA) produced a conceptual analysis very similar to the results obtained in the Gulf of Maine study, involving as key concepts **dissolved gases**, **bottom topography of the ocean**, and **water movement**.

A study of student understanding of **acidic deposition** [2] focused on a major chronic ecological crises that is slowly changing many regional environments world wide. The Northeastern USA and CN are especially affected by acidic deposition. Among the water-related concepts addressed in this subtopic were **pH**, **atmospheric water** and **aquatic ecosystems**.

Finally, another study of children's understanding focused on the complex concept of **pollution** [4]. We found that children's perspectives of concepts related to **pollution**, including **water pollution**, are molded by the cultural context in which events take place. Key concepts related to **water pollution** included the **interconnectedness of earth systems** and the **water cycle**.

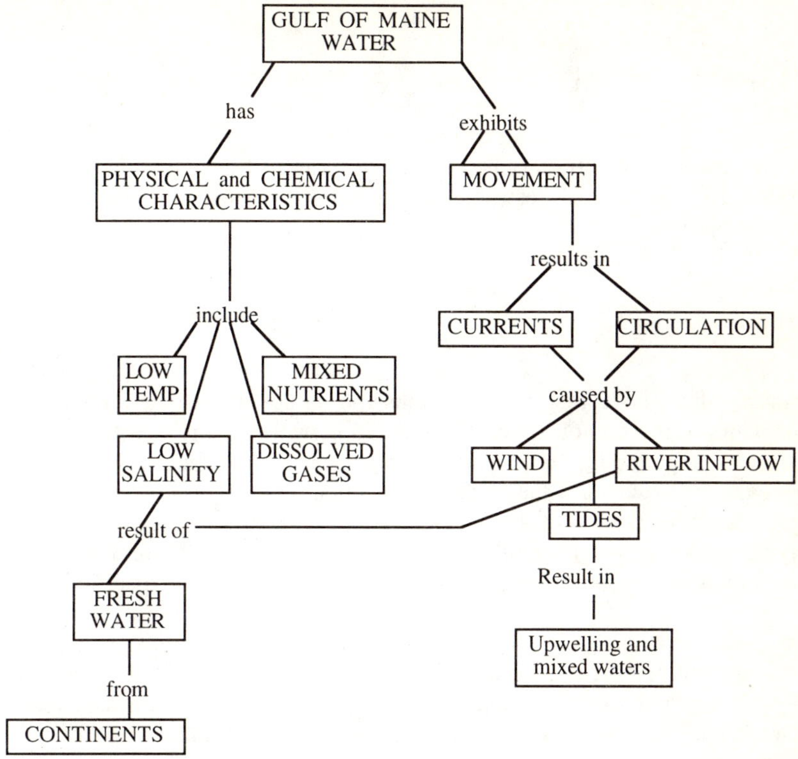

Figure 2. Concept map constructed by research team to identify critical concepts related to physical and chemical characteristics of water in the Gulf of Maine (Brody & Koch 1989-90)

These topics, which require a complex and integrated knowledge structure for understanding, have been summarized by Brody (in press).

The results of these analyses are valuable because they give a richer understanding of the nature of knowledge from numerous disciplines in relation to the concept of water. If science educators are to adopt modern paradigms for curriculum development, such as the integrated Science, Technology and Society (STS) approach or the interdisciplinary Human Ecological perspective, they must consider the complex nature of knowledge of the world in which they live. This is important for all citizens who are directly affected by science, technology and society through their environment, not just those in science-related fields. If we as educators are going to positively influence the future of people and the earth we must look for the connections between disciplines which will help illustrate the importance of interrelationships in nature.

Research Questions

This research was guided by several focal questions. First, the most general question, "What is the role of knowledge about water in understanding biology and ecology?" Second, and more specifically, "What concepts and relationships have the greatest potential for promoting meaningful learning?" Finally, "What methods are effective in restructuring knowledge about complex, interdisciplinary subject areas"?

Methods

To answer the above research questions the author constructed a number of concept maps and computer based semantic networks using SemNet software. SemNet is a general purpose intellectual tool for restructuring or creating knowledge frameworks. It facilitates construction of n-dimensional networks and allows easy modification of the net elements. It grew out of research in semantic network theory begun by Quillian (1969) and is related to previous work in concept mapping, incorporating concepts and relationships into specific instances [8].

In constructing the network, personal knowledge of the concept of water as well as several resources were used. Among the resources were university level texts focusing on biology, ecology and resource management. Primary research articles related to marine and aquatic ecology provided added insight into the concepts in the network.

Results

The author created a semantic network about water containing 593 concepts linked together by 156 different relationships to form 923 instances (concept-relation-concept units), within a period of about 20 hours (Table 1).

Table 1. About the water network

Net Elements	*# Elements*	*# Synonyms*	*# Texts*	*# Names*
Concepts	593	7	4	599
Relations	156	0	0	294
Instances	923			

One of the important measures of the integration of ideas in the net is the interconnectedness of its concepts. Table 2 summarizes the number of concepts in this net having zero, 1–2, 3 or more, etc., direct links to other concepts. Two hundred and thirty-one concepts (38.9%) participate in at least three instances each.

Table 2. Concepts with various number of instances in the water net

# Instances	*# Concepts*	*% of Total Concepts*
0	15	2.5
1–2	347	58.6
3 or more	231	38.9
6 or more	84	14.2
10 or more	33	5.6
20 or more	1	0.2

Among the most important concepts contained in the net are the ones that participate in the greatest number of instances and/or have the highest embeddedness [8]. **Water cycle** is the most developed concept with 25 instances. The most embedded concept with a measure of 150 is **biology**.

To determine which concepts are most valuable in looking at interrelationships, individual frames of the network were reviewed for interesting cross linkages (Figs. 3–6). Some of those frames illustrating how particular concepts that are related to **water** serve to link various aspects of the network are described here.

Figure 3 (next page)

The concept **polar** has some interesting properties in linking various parts of the network. The network frame illustrates the molecular to global application of water-related concepts. **Polar** draws on two major theories from distinct disciplines, electromagnetism from physics and geomagnetism from earth science. This frame from the network links geography (north and south poles), climatology (climactic regions), physics (bonding) and chemistry (solvent). As such it provides an opportunity for teachers to highlight interrelationships; however, it also presents a challenge in differentiating among various uses of the same term.

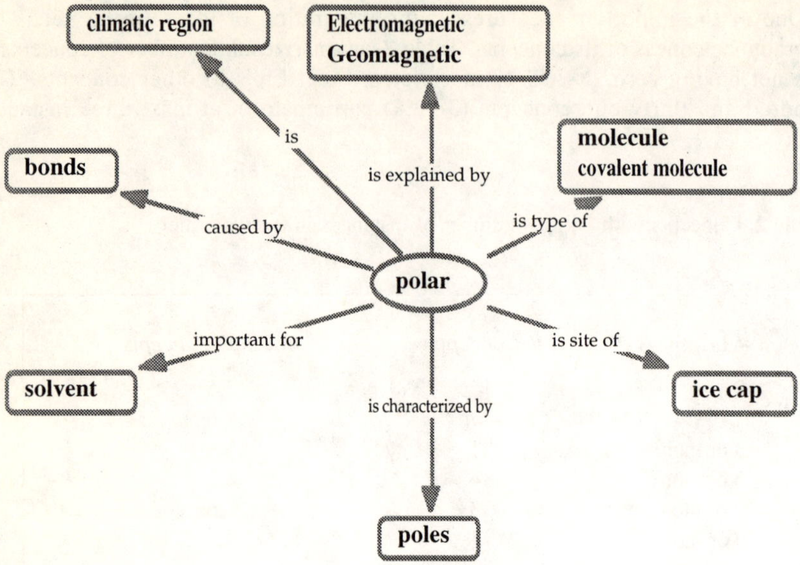

Figure 3. Section of semantic network linking several disciplines

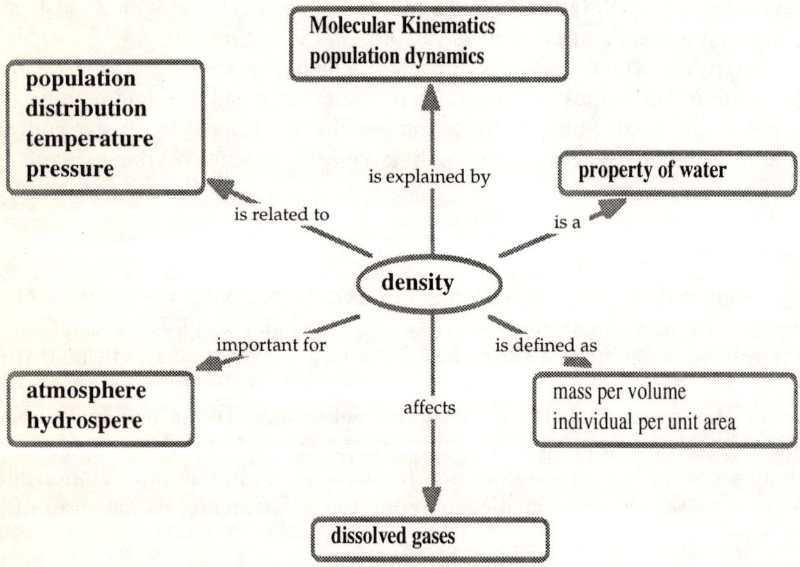

Figure 4. Frame from water net illustrating the interconnectedness of the concept of density

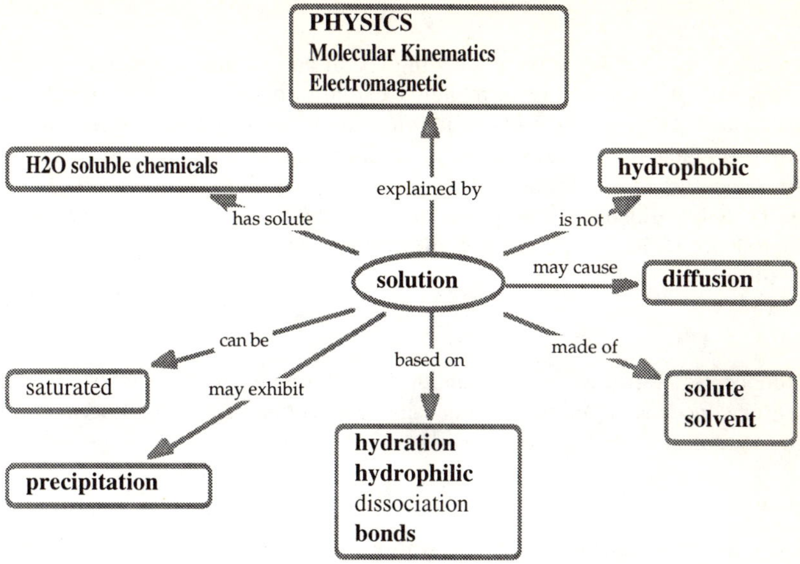

Figure 5. The concept of solution and its relationship to water

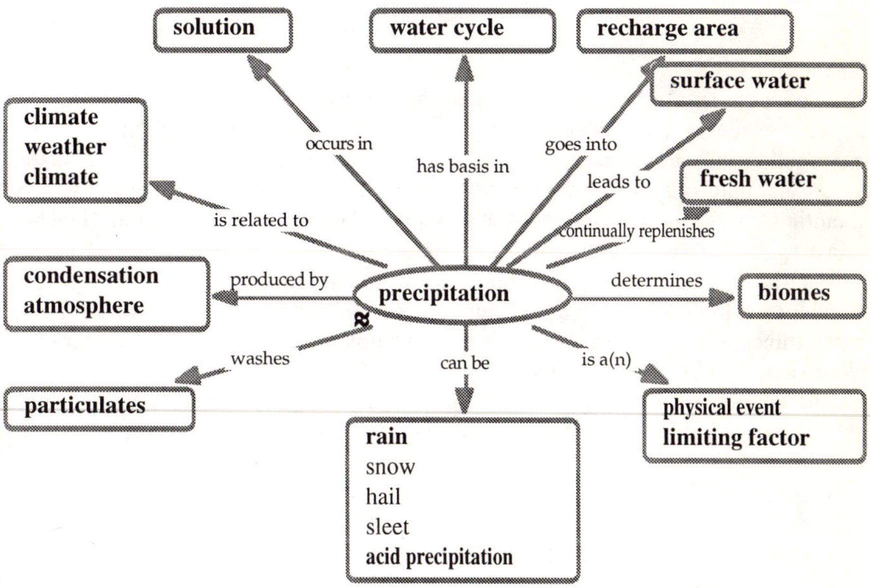

Figure 6. Water net frame illustrating the central role of precipitation and the water cycle in understanding the role of water

Figure 4

The concept of **density** links interesting social and natural science areas of the network. In one sense, the term helps explain social conditions (population distribution). On the other hand, it refers to ideas in physics (in relation to **dissolved gases**, **temperature**, etc.). **Density** is explained by two very distinct theories in this frame and is also linked to a third, **kinematics**. Finally, **density** helps illustrate how physical phenomena such as **temperature** and **pressure** are related to **properties of water**, a concept which is in turn linked to **ecology**.

Figure 5

The concept of **solution** provides another interesting example of interrelatedness of **water** concepts. Again we find the same concept name used to describe events from the molecular to the global level.

Figure 6

Finally, the important nature of the **water cycle** is illustrated through one part of the cycle, **precipitation**. **Precipitation** is linked to a range of ideas from **climate** to **weather** to **solutions** to **ecology** to **fresh water**.

Conclusions

In summary, water is fundamental in understanding biology and ecology. Among the concepts significant for understanding the roles of water are precipitation, solutions, density and polarity. One way of restructuring knowledge in ways that tie together interdisciplinary subject areas is to by focus on certain critical shared concepts that play important roles across disciplines. These ideas will be examined in terms of Schwab's four commonplaces for the curriculum: teacher, learner, curriculum and governance [13].

For the teacher, it is important to note that there are a number of critical concepts which can serve as starting points for teaching children about interconnections in nature. Water is one example of an extremely important and very rich interdisciplinary idea.

The concept of water is a familiar idea with which all learners have had some experience. Children's understanding about water ranges from correct knowledge to some interesting misconceptions. Because water is of critical concern throughout the world, it also has relevance for the learner. Water provides important links to new ideas for the learner and thus its use as a superordinate structuring concept can help promote meaningful learning.

If we are to consider new conceptions of the curriculum which help promote the interrelatedness of disciplines, we must find good examples to help us construct these meanings. Water may be a quintessential concept in terms of its interrelationships in nature.

Finally the reorientation of the curriculum towards interrelationships and the restructuring of knowledge provides an important and critical opportunity for teachers and learners to collaborate in sharing meaning about complex material. When learning and teaching focus on the complexity of nature and we all realize that no one individual can hold a monopoly on information, then team teaching and group work become the educational strategies of the future and innovative tools such as concept mapping and computer based semantic networking will play an important role in classroom instruction.

References

1. Ausubel, D.P., Novak, J.D., & Hanesian, H.: Educational Psychology: A Cognitive View, New York: Holt, Rinehart & Winston 1978.
2. Brody, M., Marion, S., & Chipman, E.: Student knowledge in fourth, eighth and eleventh grades related to acidic deposition, Journal of Environmental Education, 20 (2), Winter (1988-89).
3. Brody, M., & Koch, H.: An assessment of 4th, 8th and 11th grade students' knowledge related to marine science and natural resource issues, Journal of Environmental Education, 21 (2), Winter (1989-90).
4. Brody, M.: Fourth, eighth and eleventh grade students' understanding of pollution, Journal of Environmental Education, 22 (2) Winter (1990-91).
5. Brody, M.: A programmatic approach to understanding student knowledge of environmental issues, Proceedings of the Second International Conference on Misconceptions of Science and Mathematics, Cornell University, Ithaca, NY 1987.
6. Brody, M.: Student science knowledge related to ecological crises, International Journal of Science Education (in press).
7. Champagne, A.B., & Klopfer L.E.: Naive knowledge and science learning. Paper presented for the AAPT, 2-22,1983.
8. Fisher, K.M.: Semantic networking: the new kid on the block, Journal of College Science Teaching, 27, 10 1001-1018,1990.
9. Fisher, K.M., Faletti, J., Patterson, H., Thornton, R., Lipson, J., & Spring, C.: Computer-based concept mapping, Journal of College Science Teaching, 19 (6), 347-352, 1990.
10. Novak, J.E., & Gowin, D.B.: Learning how to learn. Cambridge, UK: Cambridge University Press 1984.
11. Novak, J.: A Theory of Education, Ithaca, NY: Cornell University Press 1977.
12. Novak, J.: Clarify with concept maps, The Science Teacher, October 1991, 45-49, 1992.
13. Quillian, M.R.: Word concepts: A theory and simulation of some basic semantic capabilities. Behavioral Sciences, 12, 410-430,1967.
14. Schwabb, J.: The practical: A language for curriculum, School Review, 78, 1-23. SemNet Software, SemNet Research Group, 1043 University Ave., Suite 215, San Diego, CA 92103, 619-232-9334, 1969.

The Information in Relations in Biology, or The Unexamined Relation Is Not Worth Having [1]

Joseph Faletti and Kathleen M. Fisher

Center for Research in Mathematics and Science Education
San Diego State University, San Diego, CA 92120-5006, USA

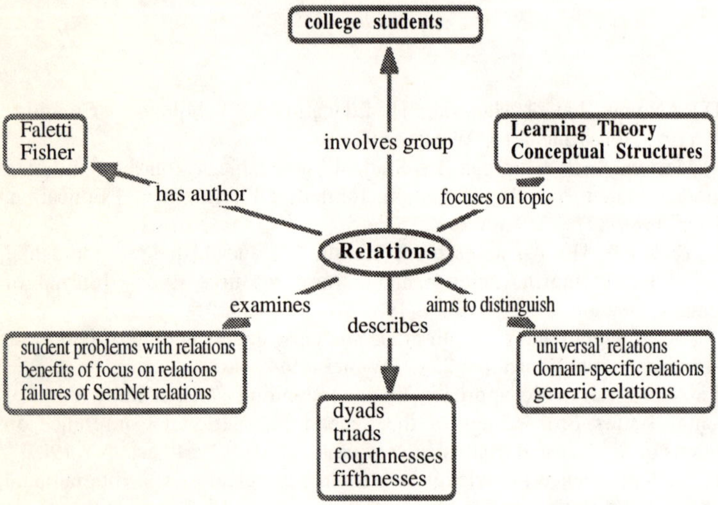

Abstract. Designing good relations is a challenge, as is using them consistently in the context of constructing knowledge representations. Further, the ability to generate and use relations effectively is a feature that clearly distinguishes between good and poor biology students. Fortunately, the SemNet software makes it possible for teachers to diagnose individual student problems in creating and using relations and to provide pencil and paper exercises to build missing skills. Assignments for students engaged in generating or using computer-based knowledge representations are significantly more powerful, in our opinion, when they are designed to prompt student thinking about using and applying knowledge to solve problems rather than to organize textbook knowledge in a relatively inert format. Finally, there are many problems in knowledge representation strategies for science students that remain to be solved.

Keywords. Biological relations, knowledge representation, concept mapping, semantic networking, SemNet, knowledge maps, cognitive analysis

---

[1] With apologies to Socrates, whose "The unexamined life is not worth living" is more elegant in 'relating' its point.

Introduction

New ideas are understood largely through connections we make between them and our prior knowledge [2, 21]. These connections are critically important in comprehension and learning. The utility of knowledge maps [14] and semantic networks [3, 6, 7, 8, 16] in promoting conceptual understanding and supporting cognitive analysis lies to a large extent in the support they provide for better analysis and specification of the nature of the relations between ideas. In many knowledge maps and networks, unfortunately, representations of relations between concepts are designated by unlabeled lines. Such unnamed connections between ideas can allow (and even promote) fuzzy thinking since they limit the communicative value of each proposition.

The first author of this paper, an educator trained in mathematics, computer science, artificial intelligence and cognitive science, has worked with several biologists including the second author and others to design SemNet™, a computer-based semantic networking tool designed to foster and support deep and meaningful learning. Faletti has been solely responsible for the software implementation. The SemNet software engages students in making their implicit knowledge explicit and provides support for individuals and groups to construct, analyze and reflect on their personal and collaborative knowledge.

In designing the software, the most difficult task has been to create suitable computer support for representing relations between concepts in a way that balances ease of use against the hard work that deep and meaningful learning requires. In the physical world we perceive relations between things (e.g., Fig. 1).

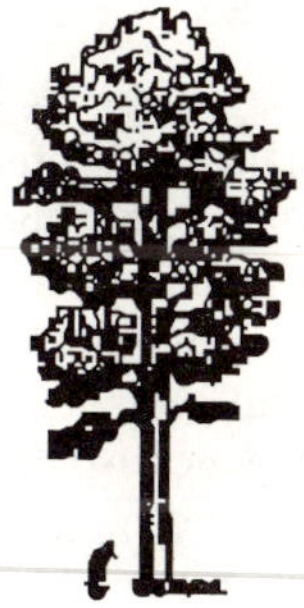

Figure 1. Fern —> *smaller than* —> Tree

Often relations perceived through imagery remain unnamed and unexamined – even when the names of the concepts they connect are familiar to us. Unexamined relations add a certain fuzziness to our thinking processes, as we often discover when we try to describe what we think we know.

Sometimes we do give names to the the relations that we perceive. The names do not necessarily capture all the nuances of the relation as we see it, but they generally capture the most critical attribute of the relation (e.g., above, below, near, hit by). Further, we rarely think of the relation in both directions (from A to B *and* from B to A); instead, we usually take a unitary perspective, looking at either how A affects B or vice versa. We are generally able to use a relatively small set of relations to connect a large set of ideas. For example, all of the billions of nuclear family units that have ever lived in Western, Christian culture can be described with a relatively small set of relations (Table 1).

Relations are generally more difficult to understand than concepts, more elusive and changeable, and more context-dependent [9, 10, 11, 12, 13]. Yet we pay very little attention to the relations that characterize each domain and subdomain we teach and learn in our classrooms. This neglect is especially true in domains that are more descriptive than mathematical in nature, as is true of most subdomains of biology.

Table 1. Relations characterizing nuclear families

First Relation Ray	*Second Relation Ray*
has child	has parent
has step-child	has step-parent
has sibling	has sibling
has step-sibling	has step-sibling
has spouse	has spouse
has former spouse	has former spouse
has significant other	has significant other

Representation of relations is challenging precisely because they are the more elusive elements in thought. As we noted above, we can recognize relations without naming them. When we do name them, we rarely think of both halves of the bidirectional connection. We generally develop, among our social or professional group, a relatively small number of standardized, domain-specific relation names. We rely heavily on general-purpose relations for communication, and these relations take on different meanings in different contexts. Thus when we give names to relations, we may not be clear about what meanings we attach to those names. And even if we are clear about our own personal meanings, the people with whom we try to communicate are likely to assign different meanings to the same names. All of these problems become apparent when a collaborative group attempts to represent their thinking about a topic in a systematic way using the SemNet software.

In this chapter we discuss ways in which the specification of relations is supported by the SemNet semantic networking software. We also examine the evolution and degree of success of particular relations used for describing biology, based on examination of hundreds of networks constructed by students, educators, and education researchers. Finally, we discuss the kinds of (usually more complex) relations that SemNet fails to support and explore their potential value.

Representing Relations for Clarity in Communication

Triads

Charles Sanders Pierce developed the first theory of relations in 1891 [19, as described in 23]. Pierce described monads, dyads, and triads (or 'firstnesses, secondnesses, and thirdnesses'). In Pierce's terminology, a monad is a concept, a dyad is a pair of concepts with the relationship between them unspecified, and a triad consists of two concepts 'mediated' by a third specialized concept which is in fact a relation. Pierce argued that there is no need "to go on to fourthness, or fifthness, and so on, because in almost every case, these higher relations can be reduced to combinations of firstness, secondness, and thirdness. On the other hand, genuine thirdness can no more be reduced to secondness than can genuine secondness to firstness."

In a SemNet-style semantic network, information is similarly represented in triads, also called 'instances' or 'propositions'. This 'atomization' seems to be satisfactory for representing a significant proportion of the ideas learned in introductory biology courses as Pierce predicted.

All relations are bidirectional and thus consist of two opposing and complementary relation rays. Consequently, each triad is also bidirectional with two opposing parts:

concept A —> *relation-ray* —> **concept B** ,
concept B —> *opposing relation-ray* —> **concept A** .

This triadic structure has the advantage of being easy to learn and use. It does not, however, easily provide for the sometimes important role of modifiers and qualifiers. Nor does it support the representation of conditional events and other situations in which three or more entities interact, as in the following examples:

The **flower** *will be* blue when the **pH of the soil** *is greater than* **7** ;
The **sphenopalatine artery** *gives* **arterial blood** to the **nasal septum** .

Representations of situations like these will be discussed later in this paper.

Generic and Domain-Specific Biology Relations

We have come to recognize that each domain and subdomain of biology is characterized by its own particular set of relations. Relations reflect the different scientific processes and ways of viewing objects that characterize different fields of study. For example, molecular geneticists and population geneticists both work in the same domain (genetics) and manipulate the same objects, such as **phenotypes** and **genotypes**, **genes** and **alleles**. Yet they study these objects from very different perspectives and rely on different methods of analysis, leading to observation of very different kinds of relations among the objects of study. Because relations can be difficult to understand, these differences make it challenging for experts in these two subfields of genetics (and in many other closely related subfields) to communicate with one another and appreciate one another's work. Instead, early molecular geneticists generally formed closer alliances with microbiologists. This seemed natural because the two groups shared many of the same perspectives and used many of the same procedures. That is, they were concerned with similar kinds of relations between things. As a consequence, they *spoke the same language*. (and also, used the same equipment).

Relations carry a lot of specific information for experts, especially the domain-specific relations. For example, the relation *originates from / origin of* (Table 2) has a particular meaning in anatomy. A full understanding of this relation requires knowing a great deal about muscles, tendons and bones and how they interact with one another. Table 2 illustrates some domain-specific relations from two different domains in biology.

Table 2. Domain-specific relations in anatomy and human reproductive physiology

Domain-Specific Relations in Anatomy [1]

First Relation Ray	*Second Relation Ray*
abducted by	abducts
originates from	origin of
insertion of	inserts on
medially rotated by	medially rotates
laterally rotated by	laterally rotates
receives parasympathetics from	supplies parasympathetics to
receives somatic efferents from	supplies somatic efferents to

Domain-Specific Relations in Reproductive Physiology [2]

First Relation Ray	*Second Relation Ray*
regulates	regulated by
synergizes with	synergizes with
degrades	degraded by
synthesizes	synthesized by
secretes	secreted by
stimulates	stimulated by
inhibits	inhibited by

[1] From a net constructed by Dr. Hugh Patterson at the University of California School of Medicine

[2] From a net constructed by an anonymous graduate student

In spite of the importance of domain-specific relations, generic relations are used in virtually all fields of study and are used more frequently than domain-specific relations. In fact, three generic relations account for more than 50% of all instances created in the biology networks we have studied during the past seven years. These generic relations can be described as follows. The meaning or essence of each relation is more important than any particular relation name and so alternative names are given for each one:

has part / is part of or *composed of / component in* ,
set has member / member of set or *includes / includes in* and
has characteristic / characteristic of or *has property / property of* .

Their high frequency of use suggests that generic relations embody the basic features we use to understand new ideas / things / events. Thus, just as when we encounter a newborn baby our first question usually is, 'Is it a boy or a girl?', so when we encounter a new idea in biology our first questions seem to be:

'What **kind** of thing is it?' (*set / membership*)
'What is it **made of** and / or what is it part of?' (*whole / part*)
'What are its **distinguishing characteristics**?' (*characteristic*)
'What does it **do**?' (*has function / function of*)

The pronounced frequency of these generic relations is illustrated in nine semantic networks constructed by undergraduate biology majors enrolled in an introductory college biology course (Table 3). The students worked individually. Inclusion relations, which consist of both *whole / part* and *set / member* links, are used in 49% of all instances created by the nine students. *Characteristic* relations, which include *has function / function of*, account for another 29% of the students' instances. Thus, the three generic relations described above are used 78% of the time by beginning students as they describe the biology they are learning. Hoffman [15] reports similar findings in his study of net construction by experts in a variety of disciplines worldwide.

The relation *has function / function of* represents only one way to indicate what a thing or event does. Action relations indicate functionality in a more direct manner, indicating the action performed by one thing on another. Examples of action relations in these nets are *absorb / absorbed by, synthesized / synthesized by, causes / caused by* and *evolved into / evolved from.* Many action relations are somewhat domain-specific. By somewhat, we mean that these relations are not found in all domains, but at the same time are not limited to a single domain. They will probably be used in a cluster of related domains. Action relations comprised nearly another 10% of students' instances in these nets.

Despite of the frequency of use of generic relations, domain-specific relations seem to play an essential role in domain descriptions. Several graduate students have been challenged to describe a particular domain using a very limited set of general-purpose relations. Yet they repeatedly find themselves unable to express

Table 3. Frequency of instances by relation type in nine student-generated semantic networks describing introductory biology

Relation Type	*Number of Instances*
Inclusion	1940
	1138
Role	263
Action	379
Temporal	150
Historical	51
TOTAL	3921

a particular idea without creating new a relation. This phenomenon is well-documented by Luoma-Overstreet & Allen [18] as they created a network describing John Anderson's book *The architecture of cognition* [1]. New relations were needed for describing each successive chapter.

Neglected Relations

It is well-known that each domain of science develops its own specialized vocabulary, the jargon of the field. This specialized language facilitates clear, concise and precise communication among experts about observations and interpretations. What is rarely noted, however, is the extent to which such specialized jargon includes names for concepts but not for relations. We analyzed the glossaries and indices of half a dozen popular introductory biology books, for example, and found that over 99% of the entries consisted of noun concepts, while less than 1% consisted of relationships (which are usually described with verbs, verb phrases, or prepositions). Thus, while different science domains are characterized by different domain-specific relations (as in Table 1), those relations' names are rarely unique (they are, as noted above, often used in multiple domains) and they are rarely highlighted in instruction in the ways that noun concepts are emphasized. [Anatomy may be exceptional in this respect.] The relations may take on domain-specific meanings but their names are generally taken from common language usage, a problem that is characteristic of concepts in certain fields such as physics as well.

Further, since relation names are usually drawn from ordinary language rather than being created as part of the unique specialized jargon of the field, and since they are not explicitly taught, they are not standardized and shared among the 'in group' in the way that concept names are agreed upon. Each individual, it seems, has his or her own preferred way to designate many relational links (as an *attribute*, a *property*, or a *characteristic*, for example, or as in *has type, has class, has example, has exemplar, includes,* and *is a* – all of which can be used to specify *set / member* relations, each with a slightly different nuance).

This lack of attention to relations and lack of standardization of relation jargon leads to idiosyncratic uses of names for describing particular kinds of connections. The consequence is often garbled communication, since the members of a group may not assign the same meanings to any particular relation name. This problem represents one of the biggest ongoing challenges among students who construct nets collaboratively. When collaborating group members take the time to carefully define each type of relation they encounter and to agree upon a single name or pair of names for each symmetric or asymmetric relation respectively, much confusion is avoided.

Is there an evolutionary advantage to the high degree of ambiguity associated with relations, or is it simply a consequence of the language processor that happened to evolve, an accident of nature that hasn't been too limiting for the species? This is an interesting question to ponder.

Level of Generality in Relations

Rosch, Mervis, Gray, Johnson, & Boyes-Braem [22] and Lakoff [17] describe basic level categories among concepts. Her categories apply to relations as well, which are a specialized form of concept. Among the features of basic level categories are that (1) they are the first level of a hierarchy of ideas to be named and understood by children, (2) the first level to enter the lexicon of the language, (3) the level at which subjects are fastest at identifying category members, and (4) the level at which most of our language and thought is organized. Among family relations, *has brother, has sister, has father, has mother, has daughter*, and *has so*n are basic level relation rays. However, one can represent knowledge at other than basic level categories and sometimes this is advantageous.

For example, *has sibling* can be used to designate all brothers and sisters as in Table 1. Likewise, *has parent* can be used to supplant *has mother* and *has father*. These more general relations simplify representation in SemNet. Since creating family trees provides a good introduction to the software for students, it is important to understand the reasons for moving to a higher level of generality.

First, three sibling relations are required when gender is specified (Table 4) whereas only one *(has sibling)* is required otherwise. Thus parsimony is the first argument for moving to a higher category in the hierarchy, even at the cost of losing gender information from the relation.

Even with these two arguments for moving to a higher level of generality, however, this may not necessarily be the optimum solution. For example, if human geneticists were using SemNet to keep family trees, they might prefer to use the standard nomenclature. Another potential solution in this case would be to alter the software so that both halves of each relation are visible when a relation ray is being selected, so that one doesn't have to rely on one's memory. Problems of this type are rarely encountered in representing the domains of biological knowledge in which we have had experience. In general, however, we find that creation of relation names for knowledge representation purposes requires thoughtful and parsimonious choice of words.

Table 4. Gender-specific relations describing siblings

First ray	*Second ray*	*Type*	*Relation between*
(1) has sister	has sister	symmetric	two female siblings
(2) has sister	has brother	asymmetric	female and male siblings
(3) has brother	has brother	symmetric	two male siblings

Table 5. Unique gender-specific relation names

First ray	*Second ray*	*Type*	*Relation between*
has sister	has sister	symmetric	two female siblings
is sister of	is brother of	asymmetric	female and male siblings
has brother	has brother	symmetric	two male siblings

Second, SemNet does not permit the same name to be used in more than one relation. In this case, the name 'has sister' is used in relations (1) and (2) and the name 'has brother' is used in relations (2) and (3). To construct a SemNet representation, relation ray names must be modified to avoid this overlap (e.g., Table 5). This is a clumsy solution that produces the unnecessary burden for the net creator of remembering and consistently using arbitrary variations in relation names to distinguish between different relations.

Spatial Conventions

A semantic network generally includes a rich web of connections from each key idea to other closely related ideas. Within these webs there are often many linear strings of information such as hierarchies, temporal flows and causal chains. Use of consistent spatial conventions can help to make this internal structure of semantic networks more useful and apparent. For example, *whole / part* and *set / member* relations usually are displayed vertically in the graphic mode, following Western conventions (Fig. 2). Any of these hierarchies, defined by a single relation (such as *has a part / is a part of)* or a group of relations with common meaning (such as *has part / part of, contains / contained in and composed of / component in)* can be extracted from the network by SemNet (Fig. 3).

Another spatial conventions we generally follow is to represent temporal events either left to right or else clockwise around the screen (the latter device is desirable where it is advantageous to include all steps or events in one frame).

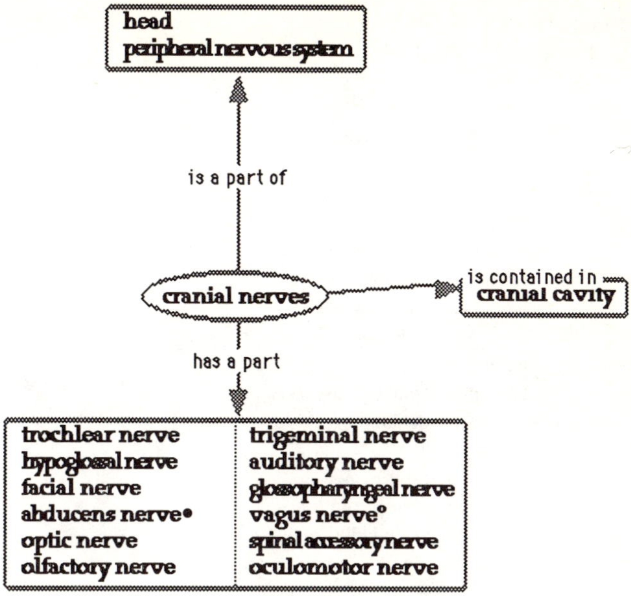

Figure 2. Vertical orientation of the *has a part / is a part of* relation in the graphic display of Patterson's human anatomy network. In the graphic display of SemNet, one central concept (in this case, **cranial nerves**) is viewed at a time with all its connections to other concepts (in this case, the relation of **cranial nerves** to 15 other concepts). By following the relation *has a part / is a part of* up and down the hierarchy, the outline in Fig. 3 is obtained.

Bi-Directional Relations

As noted above, there are two fundamentally different types of relations. In symmetric relations, the relationship between two concepts is the same in both directions (for example, **Kathy** *is near* **Ed**, **Ed** *is near* **Kathy**; **John** *has sibling* **Jane**, **Jane** *has sibling* **John**). In asymmetric relations, in contrast, the nature of the relationship between two concepts is different from each concept's point of view (e.g., **prophase** *precedes* **metaphase**, **metaphase** *follows* **prophase**; **nucleus** *contains* **chromosomes**, **chromosomes** *are contained in* **nucleus**). Most relations in biology are of the latter, asymmetric type. When constructing a large knowledge representation, it is helpful (but not essential) to use the same name in both directions as much as possible. Thus, *precedes / preceded by* might be easier to remember and use consistently than *precedes / follows*.

- **Human body**
  - **has a part head**
    - has a part cranial nerves
      - has a part trochlear nerve
      - has a part hypoglossal nerve
      - has a part facial nerve
      - has a part abducens nerve
      - has a part optic nerve
      - has a part olfactory nerve
      - has a part trigeminal nerve
      - has a part auditory nerve
      - has a part glossopharyngeal nerve
      - has a part vagus nerve
      - has a part spinal accessory nerve
      - has a part oculomotor nerve
    - has a part skull
      - has a part orbit
      - has a part cranial cavity
        - has a part middle cranial fossa
          - has a part foramen spinosum
          - has a part foramen rotundum
          - has a part foramen ovali
          - has a part foramen lacerum
          - has a part carotid canal
          - has a part superior orbital fissure
        - has a part anterior cranial fossa
          - has a part cribriform plate
          - has a part crista galli
          - has a part optic canal
        - has a part posterior cranial fossa
          - has a part internal auditory meatus
          - has a part foramen magnum
          - has a part jugular foramen
          - has a part hypoglossal canal
      - has a part skull openings
        - has a part skull foramena
          - has a part foramen rotundum
          - has a part foramen spinosum
          - has a part foramen magnum
          - has a part foramen ovali
          - has a part foramen lacerum
          - has a part mental foramen
          - has a part infraorbital foramen
          - has a part jugular foramen
          - has a part stylomastoid foramen
          - has a part supraorbital foramen
        - has a part skull fissures

Student Difficulties in Representing Relations

When students become involved in constructing systematic representations of their knowledge of biology, they find it is necessary to make the relations between ideas explicit. This appears to be a natural act for those students who actively process and try to make sense of what they are learning. However, it can be quite foreign to those who rely heavily on rote learning strategies.

Constructing relations and using them appropriately is undoubtedly the most difficult aspect of building knowledge representations. It is also the greatest source of conflict when students construct nets collaboratively. Fortunately, with this effort comes a significant payoff. Since meaning-making involves perceiving and clarifying the connections between things, we believe that knowledge representation activities provide one of the best supports for helping rote learners develop the skills they need for meaningful learning, and for helping meaningful learners polish their skills.

We have been surprised by some of the particular difficulties students have in constructing relations. SemNet networks constructed by students offer an opportunity to us as teachers to quickly diagnose each student's cognitive strengths and weaknesses in ways that have been impossible previously. These insights are especially valuable since we have found that students can make rapid progress once we have diagnosed their problem(s) and given them specific feedback and practice opportunities for building their skills.

Backwards Relations

Many lower aptitude students, especially those for whom English is a second language, do not distinguish correctly between relation rays. For example, they will often use the correct relation, such as *has part / is a part of*, but they will use it backwards, as in '**cell** *is a part of* **nucleus**, **nucleus** *has part* **cell**'. This error is so glaring it was hard at first for us to believe that it was anything more than a typographical error. However, it has established itself as a common pattern. You can imagine the difficulty students must have in making sense of biology if they cannot distinguish between the phrases that constitute the two rays of a relation. Fortunately, this is a skill they can build quickly with practice. We give written pages of exercises that lead students from simpler to more complex tasks in interpreting, orienting, and generating relations and relation rays.

Figure 3. *(facing page)* One page of the six-page hierarchy generated by following the relation ray, *has a part, / is a part of,* extracted from Patterson's human anatomy network. The entire network contains 590 concepts connected by 41 relations to form 878 instances. The network describes selected portions of the body, notably the head and lower limbs, with their innervation and blood supplies.

Inconsistent Relation Use

Another common error is to confuse *whole / part* relations with *set / member* relations. This error occurs at about the 20% level in nets constructed by experts and students alike. We encourage students to use the *whole / part* relation exclusively for physical things and their parts. However, this is difficult to achieve. One problem is that the *whole / part* metaphor is heavily used in the English language to describe all kinds of ideas. Thus, **liberty** may be part of **freedom** and **hexokinase** may be part of **glycolysis**. These are metaphoric, not literal, uses of the relation but they are pervasive. We tend to overlook some of these figurative expressions when reviewing nets, but draw the line when a student designates a **dog** as *part of* a **mammal**. Rather, a **dog** *is a type of* a **mammal** (or is *included in the set* **mammal**).

Naming the Reverse Relation Ray

A third problem is in creating relations, especially in describing a relation in the reverse direction. Each time students create a relation in SemNet, they must decide whether it is symmetric or asymmetric, and if the latter, create its name in both directions. As noted above, this is a new skill that must be acquired. Some students have difficulty at first generating the reverse names of any relations, such as those below. (Readers are encouraged to try their skills.)

has part __________________
includes type __________________
stimulates __________________

Our answers would be *part of, type included in*, and *stimulated by*. Some relations are more challenging and stump even experienced net constructors. Examples include:

has input __________________
acerbate __________________

The most common response to *has input* is *has output*, when in fact the correct response is *input to*. *Has output / output from* is a separate relation. We tell our students that the best way to be sure about choosing the correct relation ray name is to say each entire instance in both directions, either out loud or privately, as in '**glycolysis** *has input* **glucose**, **glucose** *is input to* **glycolysis**.' If they skip this step, it is very likely that they will generate mixed-up relation names.

You may notice that we include the verb to be, *is,* in a relation when it is used in a sentence (as in **glucose** *is input to* **glycolysis**), but often drop the '*is*' when the relation is used in a SemNet representation. The reason for doing this is to avoid conflicts of plurality. The relation, *is input to*, can be used comfortably only with singular concepts such as **glucose**, whereas *input to* works equally well with both singular and plural concepts such as **glucose** and **ATP molecules**.

The skeletal nature of information in a knowledge representation is one of its advantages – but teachers should be aware that this particular omission can increase the challenge for novice students who are having difficulty with relations.

Acerbate is one of those verbs that can't readily be 'turned around'. If '**John** *acerbates* **Jane**,' we don't generally say that '**Jane** *was acerbated by* **John**.' Every now and then we encounter a word like this in biology. The best solution is to find a word that is similar in meaning but better behaved. In this instance, *annoys / annoyed by* would do the trick.

Undisciplined Relation Proliferation

The fourth and last problem we'll describe is an undisciplined proliferation of relations, where many have similar or the same meanings. For example, Table 5 shows seven relations from Robert's net, all of which appear to designate a *whole/ part* linkage. Such an array of similar relations generates unnecessary and undesirable confusion.

Table 5. Apparent *whole/part* relations created by a student describing general biology (from a net consisting of 241 concepts, 29 relations, and 308 instances)

First Ray	*Second Ray*	*Frequency*
are part of	contain	15
compose	are composed of	27
are among	consist of	84
belong to	have	1
are a part of	include	11
make up	has	2
make	consists of	2

Benefits of Computer-Based Representations

Active Rather than Inert Representation

One goal of constructivist-oriented classrooms is to help students organize their knowledge in ways that are amenable for application in new contexts, rather than in inert and unproductive forms. Knowledge representation strategies can fall on either side of this fence, depending upon how such representations are implemented. We have developed a number of representational strategies that seem to enhance our students' working knowledge of biology and their ability to recall information for solving problems. Three example strategies will be described here for producing active knowledge organizations, involving (1) complex

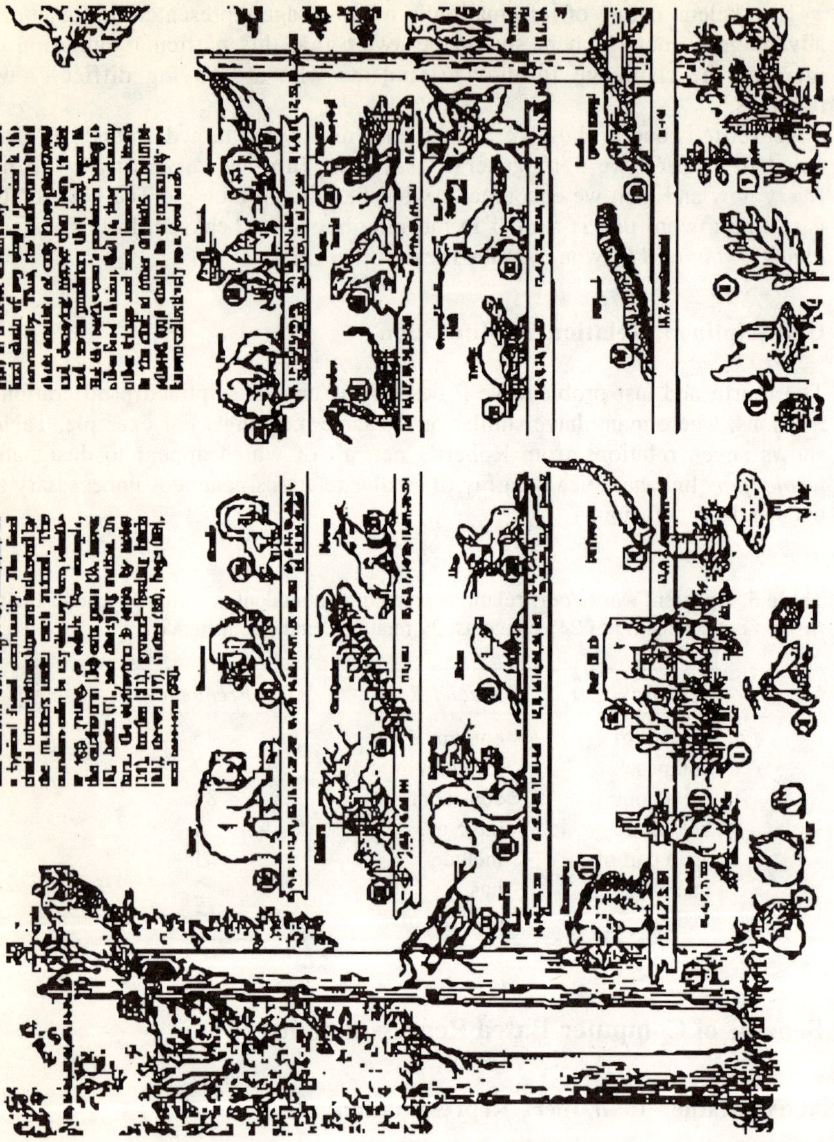

Figure 4. Forest food web

interactions, (2) coordinated functions, and (3) distinguishing characteristics. The strategies themselves are topic-specific, and so a large repertoire is desirable.

Complex Interactions (Analyzing Food Webs)

The first strategy is illustrated by a food web graphic (Fig. 4) reproduced from the National Audubon Society [20]. This graphic representation is far richer than the average food web encountered in text books, and its richness is immediately apparent. On the other hand, the details of this web are entirely obscured unless one invests a considerable amount of work, figuring out what each of the numbers represents and making lists as memory aids.

However, this food web can be represented in SemNet with quite a different effect, yielding 233 *eats / eaten by* connections among the 39 organisms. The link, *eats / eaten by,* could be described in a more elaborate scientific-sounding manner, such as *receives energy and material from / passes energy and material to*. The kind of name preferred depends in part upon the focus of the exercises (as well as individual preference).

Once the data are in SemNet, students can easily see what role(s) each organism plays in the net and which organisms they interact with (Fig. 5). This gives them the information they need to confidently classify each organism in various ways (as consumer, producer, etc., or as omnivore, herbivore, etc.).

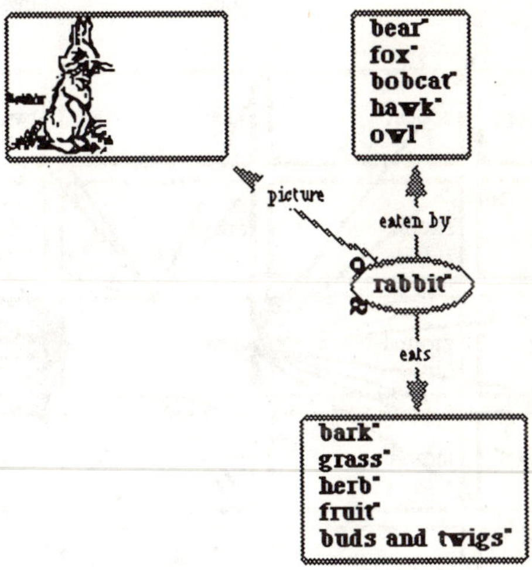

Figure 5. One of 39 frames in a SemNet representation of the food web in Fig. 4, showing all connections to the concept **rabbit**

A few relations in this food web cannot be captured in SemNet. For example, since a concept cannot be related to itself in the software, it is not possible to say that '**insects (preying)** *eat / eaten by* **insects (preying)**'. A detailed analysis of the original graphic also revealed some unlikely relations that are presumed to be errors.

Once students have transferred the information to SemNet, they can move through the network by double-clicking on related concepts and thereby trace the flows of matter and energy through the web. Students can also examine their data from various perspectives. For example, they can extract hierarchies to view all organisms ingested directly or indirectly by each individual in the web (similar to that shown in Fig. 3). The hierarchy of what the bear eats is five pages long and includes every organism in the web. They can look at their propositions in list format, review their concepts alphabetically or by their degree of embeddedness in the network, or look at overviews of all the creatures in the web such as the knowledge core in Fig. 6). The knowledge core includes up to the 49 most embedded organisms in a net spinning outward from the center in order of embeddedness. Since this net contains only 39 concepts, the knowledge core shows the entire net contents. Degree of embeddedness is calculated on the basis of the number of unique paths from each concept to two nodes away. In Fig. 6, a learner has selected the organism **insect** to view its direct connections to other organisms.

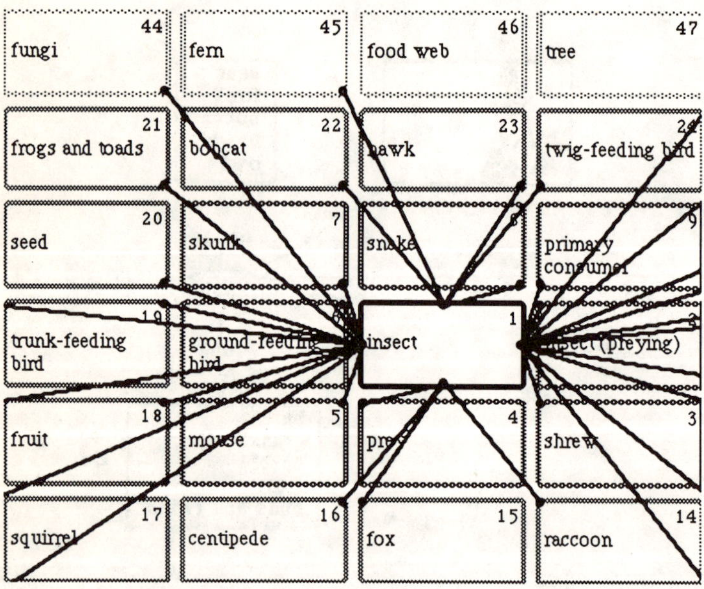

Figure 6. A portion of the knowledge core of the food web network

These varied views and the ability to shift easily among them without losing any of the 39 concepts or 233 connections provides a powerful memory extender for the students during their analyses of the food web interactions. Students can use this network to perform various analytical tasks, such as classifying each organism according to its role(s) and trophic level(s) in the net, determining the effects when one type of organism is removed from the web, and tracing the flow of radioactive or heavy metal contamination from a producer such as the grass, and so on.

Students acquire a reasonably deep understanding of many biological ideas rather quickly in doing these simple exercises. One indication of their increasing sophistication is illustrated by the number of biological terms included in essays they wrote about food webs before and after studying them (Table 6). Most of these terms weren't explicitly taught but were assimilated by students from readings and discussions in conjunction their computer work and two wet

Table 6. Biological terms used in essays written by 64 introductory biology students before and after work with the SemNet-based food web

Term	*Pre*	*Post*	*Term*	*Pre*	*Post*
Adaptation	1	11	Herbivore	4	5
Autotroph	0	4	Heterotrophs	0	2
Biomagnification	0	15	Interact/interdependence	14	29
Biosphere	0	1	Niche	1	11
Birth / death	1	1	Nutrients	5	1
Carnivore	7	9	Omnivore	2	7
Community	2	16	Order	0	13
Competitive exclusion	0	5	Overpopulation	2	0
Competitor	0	16	Participants	5	19
Complexity	2	12	Processes	9	57
Consumer	17	33	Population control	6	0
Cooperation	2	8	Population	14	5
Decaying matter	8	22	Predator / predation	69	48
Defenses	0	1	Prey	3	47
Disease	0	3	Producer	10	36
Dormancy	0	6	Resource partitioning	1	8
Eat / energy transfer	2	4	Resources	12	16
Ecological balance	1	0	Role	0	1
Ecosystem	2	10	Role / specialists	3	0
Emergent property	0	5	Fitness / natural selection	4	23
Energy flow	4	19	Saprophytes	0	5
Environment	0	8	Symbiotic	0	1
Extinction	3	6	Top predator	5	34
Food chain / pyramid	2	4	Toxin protection	0	2
Habitat	2	8	Trophic levels	0	6
Hierarchy	1	13	TOTAL	226	616

laboratory exercises, one in which they dissected owl pellets to analyze their food sources, and a second in which they simulated changes in populations over time with selective pressure.

Overall, there was an almost 3-fold increase in use of biological terms from pre- to post-essays. **Predator**, mentioned 69 times in pre-essays, was the dominant idea in students' minds prior to instruction; they paid little attention to **prey**, mentioned 3 times. After instruction students recognized **predator** and **prey** in nearly equal numbers (48 and 47 mentions respectively). Many new ideas appeared in post-essays of individual students, presumably reflecting different emphases in their thinking. For example, **autotroph** (increase from 0 to >4 mentions), **biomagnification** (0 to >15), and **competitor** and **competitive exclusion** (0 to >21). Student attention on what is happening in the web is reflected in the relatively high frequencies with which post-essay students mentioned processes (57) and interactions (29).

Coordinated Functions (Mammalian Body Systems)

Non-major students are inclined *not* to pay attention to elementary details in biology. Thus, in studying the digestive system, for example, there are always those students who believe the food is passed through the liver, the pancreas, and sometimes even the bladder and kidneys. The proportion of students who hold these and other incorrect models has declined considerably since we began requiring students to describe the path of materials through each system in SemNet, using relations such as the following:

receives foodstuff from / passes foodstuff to
receives blood from / passes blood to

Using relations such as these requires students to focus on the functioning of the system overall and to see each part in the context of its roles and interactions with other elements in the system and in the body. If there are accessory organs, use of specific relations can prompt students to clarify what they do as well, as when students are required to use the relations:

receives enzymes from / passes enzymes to
receives bile from / passes bile to

Students come away from building nets in this way with remarkably clear ideas of how each system works. They perform reasonably well on a variety of tasks such as (1) constructing networks describing a system from memory, (2) writing essays about the system (see Gorodetsky and Fisher in this book, Chap. 10), and (3) responding to multiple choice questions about the system.

Distinguishing Characteristics (Food Web)

Many students do not have a good idea of what comprises a crisp clear description or explanation in biology. For example, one exercise students perform is to describe each of the 39 organisms in the food web in such a way that it is possible to identify the organisms on the basis of the description, when their names are hidden. Initially, we found that many students would unthinkingly generate (or copy from a book) long lists of characteristics, such as 20 examples of **bird** or much obscure information about **grass**. To sharpen the exercise and make them think a little harder, we now limit students to no more than three characteristics per organism. This requires them to be clever and to focus on characteristics that set each organism apart from all others. They see the power of bringing everyday knowledge into the classroom with this exercise as well. For example, the proposition '**deer** *has exemplar* **Bambi**' is immediately informative, much more so than any biological fact they can identify.

Failures of SemNet-Style Relations

Relations in SemNet can fail to communicate well. There are advantages in systematicity and ease of net generation associated with using a parsimonious number of relations, for example, but the price of parsimony is the reduction of potentially valuable distinctions. On the other hand, a tendency toward profligacy can overwhelm. Chapter 14 in this book (Gordon) examines these points very well. Important distinctions can be missed by creators of relations, as when net constructors confuse *whole / part* and *set / member* relations, and carefully-drawn distinctions can be missed by readers and reviewers. Lack of shared relation terminology is probably the biggest source of interference with understanding. Complex cognitive models and multi-faceted linkages, such as those illustrated earlier in this chapter, can be obscured in a simplistic representational environment such as that provided by SemNet.

Limitations of Relation Names in Conveying Relation Meanings

Often distinctions implied by the chosen relation names are obvious to the creator but not obvious to their peer or instructor – and even less obvious to a person unfamiliar with the field. For this reason we encourage (or require) students to define and give examples for each relation. This is not popular with students because of the time and effort it requires, but it is very valuable, especially when a group of students first begins to work together. Definitions also allow any reader who is not sure what relationship is intended can see a more detailed explanation. This is very useful when relations are used inconsistently, making their meanings difficult to infer from the propositions in which they are embedded. Sometimes, the exact nature of the intended relationship (e.g., *supinate*) cannot be clearly conveyed with text. In such cases it is possible to illustrate the relation with an icon or image, which can help, but often the ideal solution would be to see a dynamic representation of the relationship, a feature not currently available except in prototype form.

Absence of Dynamic Elaboration

One common criticism of SemNet-based representations is their homogeneity. It could be informative, for example, to assign strengths to relation rays. Strengths could be calculated by the software and/or set by the user, and then displayed visually. Differently shaped or shaded boxes around concepts could also add interest and information. Hunt [in Clarke, 3], for example, has drawn a concept map in which each concept name is embedded in a shape of the thing it describes. Color could be used to provide added information. For example, relations of a given type could be grouped by color. However, experience with the 'Color' and 'Label' menus in the Macintosh Finder suggests that there are few cultural conventions for such features.

In addition, the spatial conventions we described earlier are not readily acquired by students. These conventions are apparently not as widespread in our cultures as educators coming from disciplines with highly hierarchical structure are inclined to believe. Unfortunately this can have serious repercussions, because when relations are constructed incorrectly, hierarchies generated with those relations do not make sense.

Third, there is clearly a need for more complex representations such as schemas or 'boxed' sets of interacting concepts to represent structures and events in a coherent way, and to define problem spaces. Ideally, a learning tool would have a variety of representational options available and students could move easily from one to another.

Absence of 'Fourthnesses' and 'Fifthnesses'

Despite Pierce's conclusions [23] to the contrary, SemNet's restriction to simple **concept** *—relation—>* **concept** units is sometimes insufficient to satisfactorily capture the information to be represented. Because SemNet permits only two concepts to be included in each proposition, clever users often encapsulize a third concept within the connecting relation, as in

stomach *passes foodstuff to* **small intestine**

It is often helpful to identify the nature of particular characteristics of an object such as size or color. However, the question arises whether the category of the characteristic should be specified in the relation, such as **chlorophyll** *—has color—>* **green**, or in the related concept, such as **chlorophyll** *—has characteristic—>* **red color**. On the one hand, using the relation *has color* and other similar relations results in a large number of relation names (violating the parsimony that some users are inclined to maintain), whereas using the relation *has characteristic* results in all characteristics being bunched together, hiding the essential distinction that this is 'color' and that is 'size' that is being talked about.

If further, the distinction must be made between red light and red objects (objects that reflect or emit red light) then the desire to say that an apple reflects red light gets reduced to either **apple** *—reflects—>* **red light** or **apple** *—reflects light—>* **red**. Note that in the first case, the kind of red that is meant is specified in the related concept, whereas in the second case, the fact that light is red is embedded in the relation.

These examples all really involve three concepts, in this case, **apple**, **light**, and **red**, which can only be 'kludged' in the SemNet scheme. The problems increase when one begins to specify conditional causalities of the form 'if W or X then Y else Z' and has to reduce them to **X** *—causes—>* **Y** and **not X** *—causes—>* **Z**. A chain of relations, such as 'the **Y chromosome** *—encodes—>* **male sex characteristics** *—in—>* **humans**', becomes lost in the milieu of a representational system such as SemNet, which recognizes only 'thirdnesses'.

Conclusion

Designing good relations is a challenge, as is using them consistently in the context of constructing knowledge representations. Further, the ability to generate and use relations effectively is a feature that clearly distinguishes between good and poor biology students. Fortunately, the SemNet software makes it possible for teachers to diagnose individual student problems in creating and using relations and to provide pencil and paper exercises to build missing skills. Assignments for students engaged in generating or using computer-based knowledge representations are significantly more powerful, in our opinion, when they are designed to prompt student thinking about using and applying knowledge to solve problems rather than to organize textbook knowledge in a relatively inert format. Finally, there are many problems in knowledge representation strategies for science students that remain to be solved.

Acknowledgments

The research reported here has been supported by the College of Sciences at San Diego State University, the San Diego State University Foundation and by grants from the National Science Foundation (TPE-9053803, USE-8951817, and TPE-8950315), the Apple Classroom of Tomorrow Program, and the State of California Lottery Funds.

References

1. Anderson, J.R. (1983). The architecture of cognition. Cambridge, MA: Harvard University Press.
2. Ausubel, D.P. (1963). The psychology of meaningful verbal learning. New York: Grune & Stratton.
3. Clarke, J.H. (1990). Patterns of thinking. Needham Heights, MA.
4. Faletti, J., & Frase, L.T. (1991, April). Automatic overviews of semantic networks from statistical and graph algorithms. Presented at the annual meeting of the National Association for Research in Science Teaching. Lake Geneva, Wisconsin.
5. Fisher. K.M. (1990). Semantic networking: The new kid on the block. Journal of Research in Science Teaching, 27(10), 1001-1018.
6. Fisher, K.M. & Faletti, J. (1989, April). Student strategies in building semantic networks in biology. Presented at the annual meeting of the American Educational Research Association, Boston.
7. Fisher, K.M., Faletti, J., Patterson, H., Thornton, R., Lipson, J., & Spring, C. (1990). Computer -based concept mapping. Journal for College Science Teaching, 19(6): 347-352
8. Fisher, K.M., Faletti, J., & Quinn, C. (1990). Exploring cognitive structure with semantic networks. SemNet Research Group Technical Report, c/o Fisher, San Diego State University.
9. Gentner, D. (1978). On relational meaning: The acquisition of verb meaning. Child Development, 49, 988-998.
10. Gentner, D. (1981). Some interesting differences between verbs and nouns. Cognition and Brain Theory, 4, 161-178.
11. Gentner, D. (1981). Verb semantic structures in memory for sentences: Evidence for componential representation. Cognitive Psychology, 13, 56-83.
12. Gentner, D. (1981). Integrating verb meanings into context. Discourse Processes, 4, 349-375.
13. Gentner, D. (1982). Why nouns are learned before verbs: Linguistic relativity versus natural partitioning. In S. Kuczaj (ed.), Language Development: Language, cognition, and culture. Hillsdale, NJ: Erlbaum.
14. Good, R.G., Novak, J., & Wandersee, J.H., (eds.) (1991). Journal of Research in Science Teaching – Special Issue: Perspectives on Concept Mapping, 27(10), 923-936.
15. Hoffman, R.P. (1991). Use of relational descriptors by experienced users of a computer-based semantic network. Unpublished Master's Thesis, San Diego State University, San Diego.
16. Jay, M., Alldredge, S., & Peters, S. (1990, April). Student semantic networks: an alternative way to present the subject. Presented at the annual meeting of the American Educational Research Association, Boston.
17. Lakoff, G. (1987). Women, fire, and dangerous things: What categories reveal about the mind. Chicago: University of Chicago Press.
18. Luoma-Overstreet, K. & Allen, B. (1990). SemNet Journal: A documentation of progress over the duration of the final assignment. Unpublished manuscript. Available from B.S. Allen, Department of Educational Technology, San Diego State University, San Diego, CA 92182.
19. Pierce, C.S. (1891). Architecture of theories. The Monist, 161-176.

20. National Audubon Society. (1965). Audubon Nature Encyclopedia, Volume 4, pp. 708-709. New York: Curtis.
21. Novak, J., & Gowin, D.B. (1984). Learning how to learn. Cambridge, UK: Cambridge University Press.
22. Rosch, E., Mervis, C., Gray, W. Johnson, D., & Boyes-Braem, P. (1976). Basic objects in natural categories. Cognitive Psychology 8, 382-439.
23. Wainer, H. (1992). Understanding graphs and tables. Educational Researcher 21(1), 14-23.

Eliciting and Representing Biology Knowledge with Conceptual Graph Structures

Sallie E. Gordon

Department of Psychology, University of Idaho, Moscow, ID 83843, USA

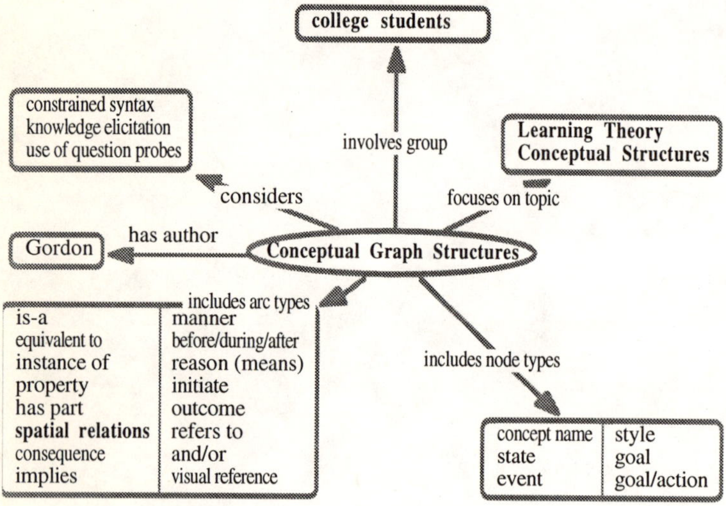

Abstract. This chapter describes recent work on methods for studying the organization and use of knowledge by making knowledge structures explicit in graph form. In the procedures, we use question probes and other complementary methods to elicit knowledge from novices and/or experts. The information is then represented in network form using conceptual graph structures. Graphing knowledge using this syntax allows one to do a number of things, including, identify student misconceptions or gaps in knowledge structures, predict student problem solving performance, allow students to graphically depict their own problem solving structures, and edit expert graphs for instructional design purposes.

Keywords. Knowledge representation, knowledge elicitation, conceptual structures, assessing, misconceptions, question probes, instructional design

Introduction

As evidenced by the content of this book, the organization and use of scientific knowledge, including biology, is increasingly becoming a focal point for educational researchers. The underlying assumption is that if we can understand how students acquire, organize, and use knowledge structures, we can then begin to modify instructional strategies to support more efficient and effective learning.

My research goals at the University of Idaho have been to develop and evaluate methods to elicit knowledge structures and study their use in problem solving [1, 2, 3, 4, 5]. At the start of this work, we realized the need to adopt some type of knowledge representation format or 'medium' to provide the backbone for our work. After identifying and subsequently modifying an appropriate representational format, we spent several years developing and testing methods for eliciting and representing both novice and expert knowledge structures. The goal of this chapter is to briefly present the methods and results of these research efforts and suggest their applicability to biology education. More specifically, I will describe:

(1) The formal graphical syntax, termed *conceptual graph structures*, that was adopted to represent knowledge in a variety of domains;
(2) The methods developed to elicit knowledge structures and study their use in problem solving;
(3) Methods developed for knowledge engineering in instructional design, the process of improving completeness and semantic coherence in text and other instructional systems; and
(4) A summary of the potential application of these methods to biology instruction.

While this work has been conducted in domains other than biology, I hope that readers will see the direct applicability of the methods to the biological sciences.

Conceptual Graph Structures: A Formal Syntax for Representing Knowledge

Formats for Representing Knowledge

Many educational researchers have found it useful to adopt a network representation format for explicitly representing knowledge structures [e.g., 6, 7, 8, 9]. More specifically, many have chosen to use *semantic networks*, which have found a wide range of application in science education [e.g., 6, 8, 10]. In this work, semantic networks consist of nodes connected by links, where each node contains a single unitary concept and each link is a single relation between two nodes. The content of nodes and links is open, being left to the researcher, teacher, or student to designate, modify or extend.

Semantic networks are simple and flexible making them attractive for many purposes. However, there are also certain drawbacks, and these drawbacks have caused many researchers in cognitive science to move towards other more powerful representational methods as long as fifteen years ago [e.g., 11, 12, 13]. For the purpose of this work, several problems with the use of simple semantic networks were identified. First, they are almost always developed using relatively unconstrained graph syntaxes. Other than specifying that 'concepts' belong in graph nodes, and 'relations' are used to label graph links, researchers, teachers, and students are usually free to map the knowledge in whatever way seems appropriate at the time. This flexibility makes it difficult to standardize graphs and compare or analyze them in any meaningful way.

Second, there are other representational syntaxes that can more easily accommodate the representation of relationships between complex structures, such as propositions. In particular, propositional networks or conceptual graphs have been used by cognitive scientists to simulate the acquisition, organization, and use of complex knowledge structures [e.g., 12, 13, 14, 15, 16, 17]. One reason propositional networks have proved useful is that they have the ability to represent propositions *as well as* single concepts within one node (recall that semantic networks generally use single concepts for each node). This gives propositional networks the ability to easily represent the relationship between two or more propositions, an important component in complex human thought.

For the reasons outlined above, we chose to adopt a formal and constrained propositional graph syntax that can capture complex knowledge and its relationships. Specifically, we chose to use *conceptual graph structures* a formal and constrained graphical syntax developed and tested by Art Graesser and colleagues to study the representation of knowledge during prose comprehension [12, 14, 18, 19, 20, 21, 22, 23]. This syntax has been used for over ten years and has received extensive validation through empirical research and computer simulation [e.g., 19]. The syntax has been shown to be applicable to a wide variety of domains including both everyday 'world' knowledge and more expository types of information such as that found in the sciences [20]. For our work on knowledge acquisition, we modified Graesser's representational format slightly, and it is this modified syntax that is summarized here (see [4] for a comparison of the two syntaxes).

Components of Conceptual Graph Structures

Conceptual graph structures, or conceptual graphs, consist of nodes that are linked by labeled, directional arcs. Figure 1 shows an example of a small, incomplete graph for information related to photosynthesis. Each node in the graph specifies the *type* of information found in the node (e.g., concept, state, etc.) and also the *specific* information itself (e.g., 'pigment' or 'pigments absorb photons'). Each node in the graph can be either a concept or statement, and is categorized as one of the following types:

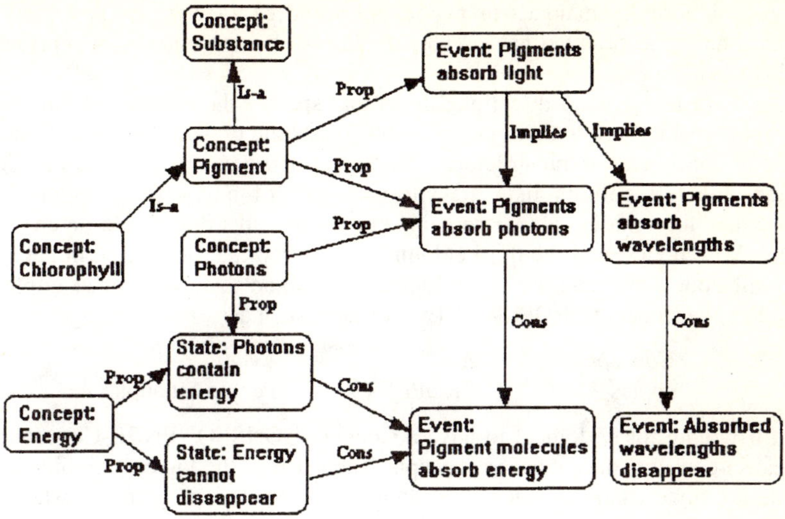

Figure 1. Part of conceptual graph structure for photosynthesis. Arc abbreviations are: Cons = Consequence, Prop = Property

- Concept name
- State (statement)
- Event (statement)
- Style (statement)
- Goal (statement)
- Goal/Action (statement)

A *concept* is a concrete or abstract discrete entity such as 'tree', 'cell', 'primary cell wall', or 'genetic origin'. *States* are conditions that 'remain unchanged throughout some time frame' [14] and are typically described with phrases. Examples would include 'plant cells have walls' and 'animal cells do not have cell walls.' An *event* represents a state change within some time frame, such as 'cells secrete hardening substances into the primary wall' and 'human muscle cells make ATP.'

Style is a special category of information that denotes the specific way in which an event happens or a goal/action is carried out, as in 'the style of the molecule movement is random.' A *goal* node specifies some state or event desired by a person or organism, such as 'body to be warm' or 'computer to be off', that may or may not end up being achieved. Finally, a *goal/action* node specifies an action or activity that is also a goal, such as 'put on a coat'. Goal and goal/action nodes are not common elements in conceptual graph structures representing factual information such as that found in biology text. They are very frequently found in representations of knowledge having a procedural orientation such as biology laboratory manuals or mathematics and engineering texts. Given their intentional, goal-oriented nature, they could also be appropriate for representing animal behavior.

Arcs are labeled and directional links that specify the nature and direction of the relationship between two nodes. Table 1 lists the predominant nodes and arcs used in conceptual graph structures. The nodes and arcs are arranged in columns where the left column lists the various types of nodes from which an arc emanates, the middle column contains the various arcs that would be emanating from those nodes, and the third column lists appropriate or legal terminal nodes. Not all node types and arc types logically combine together. For example, one would have Concept-PROPERTY-Event triads, such as:

Concept:	*Arc*	*Event*
apples	PROPERTY	People eat apples

But it would not be logical to have a Concept-CONSEQUENCE-Concept triad, because concepts do not cause concepts. Table 1 lists the logical combinations of node-arc-node triads, divided according to the type of subgraph in which they are typically found. Subgraphs and the various arcs listed in Table 1 will be described further in the following section.

Graph Subtypes

Conceptual graph relations (arcs) tend to cluster into certain graph subtypes or substructures. Knowledge of these substructures can make it easier to understand and analyze the organization of scientific text as well as the knowledge structures of our students. To illustrate the different types of subgraphs, Fig. 2 shows a graph for some knowledge relevant to the use of a hi-fi VCR with a jog-shuttle control. This graph contains within it all of the subgraphs described below. As each subgraph and its associated arcs are described, the reader is also referred back to Table 1 which lists the appropriate node-arc-node combinations.

One of the most common types of subgraph is the *taxonomic* structure. Taxonomic structures convey the meaning of concepts, predominantly through *Is-A* and *Property* arcs. For a physical system, the taxonomic structure specifies the components, their taxonomic relationship, their partonomic relationships (the parts of a system or component), etc. In Fig. 2, nodes 10–11, and 15–17 represent taxonomic subgraphs. In some taxonomic graphs, it is desirable to include specific instances of the concepts, events, and states. These are tied into the graph via *Instance-of* arcs. The taxonomic structure is the graph type most commonly used in educational research (and, in fact, sometimes the *only* type).

Other subgraph types include spatial, causal, and goal hierarchy structures. *Spatial* structures specify the spatial layout of regions and objects in regions. For spatial substructures, the arcs are sometimes domain-specific, as in ‘Washington *Is North of* Oregon’ or ‘adipose tissue *Is Below* skin.’ In Fig. 2, nodes 11–15, 18, and 19 show a spatial subgraph representing physical layout of the system.

Table 1. Node and Arc Combinations for Different Subgraph Types

Source Node	Arc Emanating From Source Node	Terminal Node
TAXONOMIC STRUCTURE:		
Concept	IS-A	Concept
Concept, Event, State	EQUIVALENT-TO	[similar node type]
Concept, Event, State	INSTANCE-OF	[similar node type]
Concept	PROPERTY	Event, State, Goal, Goal/Action
Concept	HAS PART	Concept
SPATIAL STRUCTURE:		
Concept	SPATIAL RELATION e.g., Contains, Left-of, North-of, etc.	Concept
CAUSAL STRUCTURE:		
Event, State, Style	CONSEQUENCE	Event, State, Style
Event, State, Style	IMPLIES	Event, State, Style
Event, Style	MANNER	Style
Event, State	BEFORE/DURING/AFTER	Event, State
GOAL HIERARCHY:		
Goal, Goal/Action	REASON (or MEANS)	Goal, Goal/Action
Event, State, Style	INITIATE	Goal, Goal/Action
Goal, Goal/Action	BEFORE/DURING/AFTER	Goal, Goal/Action
Goal/Action	MANNER	Goal/Action, Style
Goal/Action	CONSEQUENCE	Event, State, Style
Goal, Goal/Action	OUTCOME	Event, State
GENERIC ARCS:		
Goal, Goal/Action, Event, State, Style	REFERS-TO	Concept
Goal, Goal/Action, Event, State, Style	AND/OR	Goal, Goal/Action, Event, State, Style
Goal, Goal/Action, Concept, Event, State, Style	VISUAL-REF	[visual image or image label]

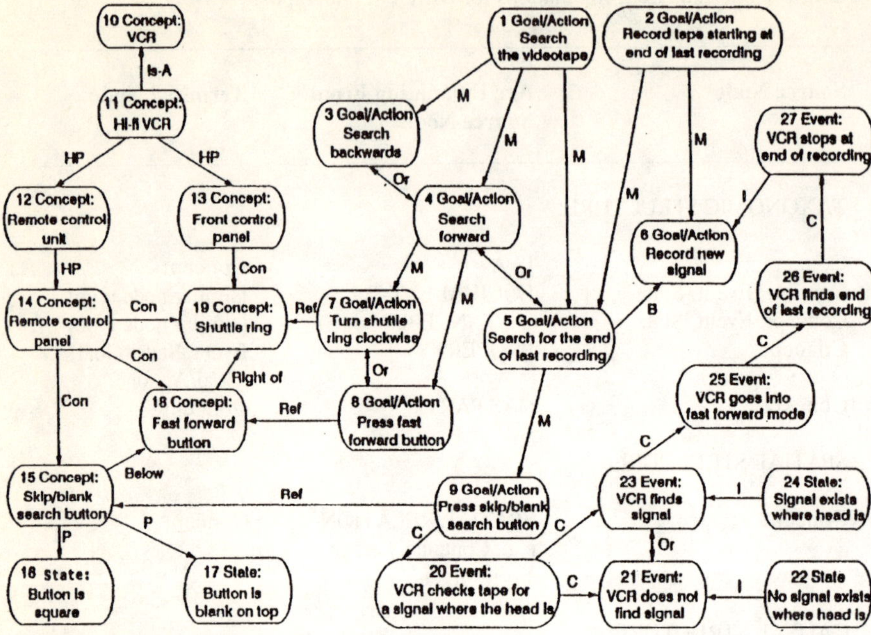

Figure 2. Partial conceptual graph for knowledge about a hi-fi VCR. The abbreviations for arc categories are: C = Consequence, Cont = Contains, HP = Has Part, I = Initiate, M = Means, P = Property, Ref = Refers to

Causal structures represent the causally driven state and event chains in physical, biological, and technological systems. The predominant type of arc in causal structures is the *Consequence* arc, which represents the entire class of relations including *causes*, *is followed by*, *has the consequence*, etc. The causal structure in Fig. 2 is a chain moving from node 20 up through 27.

Goal hierarchies represent planned action sequences executed by animate agents. They are usually clusters of goal/action nodes interconnected either by upwardly directed *Reason* arcs or by downwardly directed *Means* arcs (these are complementary arcs). Other important arcs in goal hierarchies include *Initiate*, where some state or event initiates one or more goals and activities in the goal hierarchy, and *Outcomes*, where various states and events are outcomes of the goal-oriented activities. The goal hierarchy in Fig. 2 is represented by links between nodes 1–8.

In addition to the arcs described above, there are certain types of links commonly found in all types of subgraphs, *Refers-to*, *Visual-rep*, *And* and *Or*. The *Refers-to* arc is a referential arc indicating that a part of one node is a concept referenced in another node. For example, 'density gradient *Refers-to* density.' Inclusion of all *Refers-to* arcs will expand a graph very rapidly, and is quite optional depending on the circumstances. In cases where it is desirable to have visual information represented in a graph, this information can be represented as nodes linked to relevant text node(s) via *Visual-rep* arcs.

While we usually specify only one arc between a pair of nodes, we also recognize that there is always implicitly a complementary arc, pointing in the opposite direction (e.g., *Has Part* has a complementary arc of *Is Part of*). In addition, we might add one or two domain-specific arcs in a given area when the domain-general arcs are too cumbersome (e.g., *Has equation* in a mathematics domain is much easier than trying to use domain-general arcs to represent the same idea).

Some Advantages of Using a Constrained Graph Syntax

Conceptual graph structures are composed by using a very specific and highly constrained graph syntax. As noted in the previous section, it has specific 'legal' terms and relationships that clearly take a certain period of time to learn. Even so, there are certain advantages to using such a syntax that will be briefly noted in this section.

Some of the advantages of using graphs of *any* kind for knowledge representation have been outlined in previous publications [e.g., 4] and will be summarized here. First, sentences may contain ambiguities that make the meaning of the sentence unclear, and such ambiguities are much less likely to be found using a formal graphical syntax (one reason they are used in artificial intelligence research or AI). Second, graphs act as a good shorthand notation method during interviews where knowledge acquisition takes place. Third, graphs contain explicit relational information between concepts and statements that is usually missing in sentence formats. Fourth, graphs visually show the underlying organization of information or lack thereof.

There are also certain reasons for using a constrained graph syntax rather than simply writing whatever nodes and links one wishes. First, from the standpoint of efficiency, it makes sense that, as information processors, people would have a constrained set of relationships that can be used across domains. Once we learn the meaning of basic relationships such as 'is part of', 'has the consequence of', 'is a characteristic of', etc., we can use that knowledge as the basis for very powerful thinking and reasoning skills. That is, we can hypothesize that a generic set of relations forms a template that can be applied to each new domain. After the application of this template, a variety of inferential reasoning skills based upon those relationships can be transferred to the domain. The parsimony and efficiency of such a wide application of inference rules has been demonstrated in recent complex knowledge bases developed in AI [e.g., 13]. If learning in a new domain required acquisition of new relations as well as new concepts, one implication is that we would have to acquire a completely new set of basic level inferential and reasoning skills as well. For interested readers, Graesser has given a variety of theoretical grounds for using the conceptual graph structure syntax [19, 22].

There are also several pragmatic reasons for use of a relatively constrained and domain-general syntax. First, the use of a domain-general and highly constrained graph syntax is like learning a language. Once learned, it is easily used in a variety of applications without having to learn a new language. That is, after a person becomes familiar with the syntax, it is easier to identify the nature of a specific relation than if one had a virtually unlimited choice.

The second pragmatic advantage is that the use of a constrained syntax has led to a particular knowledge acquisition method, *question probes*, that has proved to be highly effective in eliciting knowledge in a wide variety of domains [1, 2, 4, 24]. Use of the question probe method (described later) requires that information be represented in conceptual graph format. And third, it is easier to inspect and directly compare graphs when they are based on a specific and constrained syntax for nodes and links.

As a final note, we have pursued several projects revolving around the explicit identification of student or 'learner' knowledge structures, in a variety of domains. However, none of these activities has required the learners to graph their own domain knowledge. Thus we were not concerned about an uninitiated learner's ability to use a particular graph syntax, and cannot therefore comment on the ease or difficulty with which students might use the syntax.

Methods for Studying the Organization and Use of Knowledge Structures

Mapping Knowledge Structures with Question Probes and Conceptual Graph Structures

We use conceptual graph structures to explicitly represent a person's domain knowledge at a very detailed level, as opposed to concept mapping techniques that result in a more high-level skeletal structure [e.g., 6, 7, 8]. We then evaluate how students use this knowledge in subsequent problem solving activities. Clearly, the first critical question is how to elicit and map an individual's knowledge structure(s) such that the map bears some resemblance to the actual knowledge structure(s).

In a series of studies, we developed and tested a methodology for explicitly representing individuals' domain knowledge structures [1, 2, 4]. This methodology consists of asking very specific question probes regarding the particular topic. Answers to these questions are then translated into conceptual graph structures.

Such question probes are designed to elicit information in all of the substructure areas described above: taxonomic, spatial, causal, and goal hierarchies. They consist of questions about specific knowledge, such as "What is genetic recombination?" or "How does gel electrophoresis separate macromolecules?" Question probes are designed to be very specific and cover the knowledge base in detail. They are generated in a systematic fashion, a process which can be done in one of two ways.

The first method for generating question probes is to start with instructional information on a particular topic. For example, we developed instructional videotapes and then translated them into conceptual graph structures [2]. The instructional graph was then used as the basis for developing question probes to administer to learners. More specifically, each node on the graph formed the basis for a set of questions (a process to be described shortly).

The second approach does not require an instructional graph to act as the basis for question probes. In this case, the student is asked very general questions at first, such as “Tell me about cells.” This interview process is termed ‘free generation’ [25]. The information learners give is then translated into conceptual graph form, and the nodes in the graph are used as a basis for additional question probes. One danger in taking this approach is that novices often have relatively fragmented knowledge structures [2]. As a result, using the subject’s own starting point is likely to leave some knowledge structures untapped. However, this method is quite appropriate for those instances where one is interviewing an expert and mapping the structure for instructional purposes.

Regardless of which of the two methods are employed, question probes are generated on the basis of conceptual graph contents. One or more probes were generated for each node on the conceptual graph. To determine the probes, we developed a set of generic question probes that are appropriate for each node type. For example, for a *concept* node, the generic question probes are:

What is ______?
What are the types of ______?
What are instances of _______?
What are the properties of ______?
What are the parts of ______?

The sets of generic question probes for the remaining types of nodes (state, event, etc.) are specified in [4]. To determine the question probes to be used, the appropriate set of generic probes for a given node type is combined with the information contained in that particular node. Thus, if one concept node in the instructional graph is ‘cell’, the generic question probes would be combined with the concept of cell to yield:

What is a cell?
What are the types of cells?
What are some instances of a cell?
What are the properties or characteristics of cells?
What are the parts of a cell?

The learner is asked question probes for all nodes in the graph (either the instructional graph or the one developed through free generation).

Information obtained as a response to the probes is translated into conceptual graph form. For example, if the student states that "Meiosis occurs in the sporangia, and haploid spores are released," this knowledge would be represented as the structure shown in Fig. 3. This translation can be done as the answers are verbalized, or alternatively, answers can be tape-recorded and graphed later.

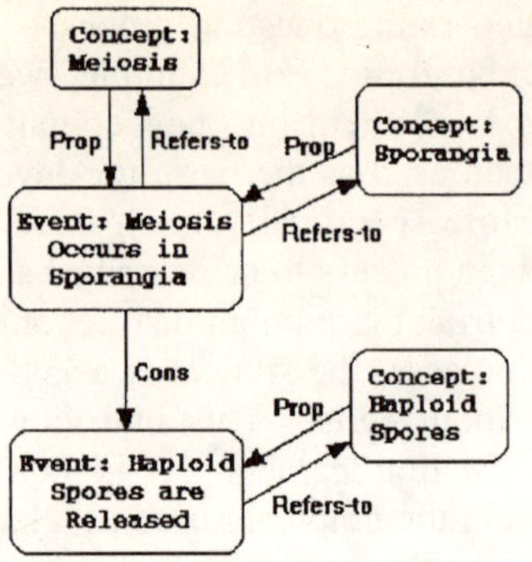

Figure 3. Conceptual graph segment for "meiosis." Arc abbreviations are: Cons = Consequence, Prop = Property

The processes described above will generally yield very large graphs with hundreds of nodes. We usually subdivide these and draw them on large sheets of drafting vellum. The advantage of this is that large areas of the graph can be seen at one time. This makes it much easier to evaluate them. However, the disadvantage is that changing the graphs can be very time-consuming. An alternative is to use a computer program such as SemNet [6] to develop and store the graphs. The disadvantage is that only a small portion of the graph can be viewed at one time, and the software does not currently have the capability to print the entire graph.

Empirical Support

We have validated the use of question probes and conceptual graph structures to elicit knowledge structures in two different ways [2]. First, we conducted a study where subjects watched an instructional videotape on the functioning and use of a hi-fi VCR. We asked them to provide a free recall protocol for the instructional material, and also administered to them a set of question probes that corresponded to the information in the instructional tape. We did not ask questions that would have caused new inferences to be made. Cluster analysis was performed on each person's free recall protocol, and a subsequent analysis of the data showed a very high relationship between the clustering of concepts in individual conceptual graph structures and the clustering of concepts in that same individual's free recall protocol. Given that cluster analysis is a fairly well-accepted indication of cognitive organization [26], this provided convergent validity for the question probe and conceptual graph method.

In a second study, we showed instructional videotapes to college students in two different domains, the hi-fi videocassette recorder (VCR) and engineering mechanics [2]. After a break, subjects received question probes, a second break, and then were asked to solve several problems that required use of the instructional information as well as small to moderate amounts of inferential reasoning. Judges blind to the problem solving performance of the subjects were asked to use the contents of the individual graphs to predict each subtask activity for each individual subject. These predictions agreed with the subtask performance (regardless of whether the subtask performance was correct or not) on 87% and 93% of the subtask activities, for the engineering mechanics and VCR domains, respectively. This lends validity to the accuracy of the contents of the graphs.

In using the conceptual graph structures to evaluate individual problem solving, we found that even when subjects held the correct knowledge required for task performance, they were not able to use it if successful problem solving required more than a few simple inferences based upon that knowledge. That is, almost all of the successful problem solving activity was determined by knowledge retrieval and application, rather than inferential problem solving with retrieved knowledge [2].

We also found direct or indirect evidence for several other phenomena:

(1) Problem solving seems to partly be a function of the amount of knowledge that must be held in working memory as well as the number of inferences that must be built upon that knowledge to generate a problem solving operator.
(2) Students often make simple inferences (such as the meaning of the TV/VCR switch) and avoid construction of more elaborate but correct knowledge structures based on instructional material.
(3) Students have great difficulty using semantic, factual knowledge to construct goal/action knowledge [2]. That is, they have trouble making necessary inferences to translate taxonomic, spatial, and causal knowledge into goal hierarchy knowledge.

We feel that this last factor is especially critical because many texts tend to segregate these types of information and assume the student can integrate them.

Knowledge Engineering Instructional Materials

As a result of recent successes with eliciting knowledge structures via question probes and conceptual graph structures, we concluded that systematic elicitation of expert knowledge might allow us to create graphs that could then serve as the basis for developing instructional materials. That is, these methods would allow us to *knowledge engineer* instructional materials, making them more complete and semantically coherent. This knowledge engineering process, termed *conceptual graph analysis*, is currently our major research activity.

Knowledge Engineering via Conceptual Graph Analysis

Before describing the conceptual graph analysis method, I will briefly outline our view of knowledge structures in general. This information is necessary to understand the basic rationale for the methods.

We support the distinction made by many others between *declarative* knowledge and *procedural* (perceptual, automatic, skill-based) knowledge [e.g., 15, 27, 28]. This distinction is most often drawn as a difference between controlled, knowledge-based processing and perceptual, automatic, skill-based processing [28, 29, 30, 31, 32, 33, 34]. Many researchers consider declarative knowledge to be the explicit knowledge we can verbalize, while procedural knowledge is implicit or perceptual knowledge that cannot be directly verbalized (see [29] for a review of this issue).

The relevance of this distinction to knowledge acquisition is the finding that people often have great difficulty verbalizing the *implicit* procedural type of knowledge [35, 36, 37, 38]. This difficulty translates into difficulties in the knowledge elicitation/acquisition process; a designer will often need to use one set of elicitation methods to identify the expert's explicit declarative knowledge and another set of methods to identify implicit procedural knowledge, e.g., [29]. We concluded that a variety of knowledge acquisition methods must often be used to identify the factual, rule, and implicit processing used by experts.

While the conceptual graph analysis methods were originally designed for expert system design, they also carry over into instructional design. Conceptual graph analysis consists of creating conceptual graphs of the domain material and then using question probes and other techniques with one or more experts to modify the graphs. The methods currently being used to perform the analysis are: document analysis and/or free generation, question probes, and observation with inductive analysis [3, 4, 5]. All activities are carried out by a knowledge engineer or instructional designer trained in using conceptual graph structure syntax. The procedures are usually carried out in the following order, although the knowledge engineer may sometimes have to cycle back through previous stages if the topic is particularly complex or large.

Document Analysis. In document analysis, relevant domain knowledge is identified in existing texts or articles. The information in the document is translated into conceptual graph form at a molecular level, one statement or proposition at a time. As designers translate text into graphic form, they typically notice many instances where information is missing, inconsistent, or simply ambiguous. Document analysis therefore results in graphs with sections that must be clarified through questioning of one or more experts.

Free Generation. As an alternative to document analysis, the designer may initialize a graph using the *free generation* technique [25, 40, 4] by simply asking an expert to describe basic information such as "Tell me about this topic" or "What is photosynthesis?" This method is particularly effective for obtaining the high-level, macrostructure nodes for a goal hierarchy. Even when documentation does exist and has been analyzed, experts often prefer to begin with a general overview based on their own knowledge. In this case, free generation is performed in addition to document analysis.

For free generation, the expert is not given specific questions, but is simply asked to describe the concept, system, or task. The information is tape-recorded and thus becomes a verbal protocol to be analyzed. This information is then translated into conceptual graph form as was done for document analysis. If a document analysis has already been conducted, the two sets of information can be integrated into one graph. Any inconsistencies can be clarified through subsequent interviews.

Question Probes. After initialization of the graph based on documentation analysis or free generation, the graph is expanded and clarified through the use of question probes [4]. For each node on the graph, the expert is administered the corresponding set of question probes, as described previously. Answers are sketched onto a graph and/or tape-recorded so that the information can be added to the graph at a later time. This process is pursued iteratively to continue expanding and organizing the graph until each section has reached the domain boundary, as judged by the expert being questioned [4].

In addition to the question probes described earlier, an additional type of questioning has also proved necessary. When analyzing existing documents or verbal protocols, the instructional designer will notice ambiguities: Either statements that have multiple possible meanings or information that seems to be implied but not directly stated. These phenomena are noted on the graph and during question probe interviews, the expert is asked for clarification – e.g., "Does X contain both A and B, or only A?"

Observation/Induction. The previous methods elicit verbalizable fact and rule knowledge. However, expertise often also involves implicit procedural knowledge, such as recognition of complex visual patterns. We have found that the use of implicit knowledge often becomes apparent during question probes. For example, when asked "How do you X?" the expert will be unable to articulate an answer, or the expert will obviously hesitate and slowly generate a plausible answer. This phenomenon indicates the existence of implicit knowledge which should be evaluated using observation/induction.

A common and useful way to capture implicit skill-based performance is to ask the expert to perform the task for a wide range of task instances (as in a factorial design). For each trial, characteristics of the task and associated stimuli as perceived by the observer are recorded along with the expert response performance for each subtask. At this point the instructional designer has two choices. First, the stimuli-response sets can be indexed into the conceptual graph at appropriate points and simply taught directly. That is, students are shown the stimuli directly as examples and non-examples.

Alternatively, the knowledge engineer can work with the expert to induce the rules underlying task performance, thus making the implicit knowledge *explicit.* However, the instructional designer should be aware that they are *inductions* and thus may not represent the true basis of the behavior. In addition, designers should keep in mind that some types of highly complex implicit knowledge, such as language production, are simply not amenable to inductive analysis. If it proves possible to induce the rules underlying behavior (as indicated by successful prediction of additional behavior), this information can then be added to the graphs as goal/rule structures and the examples tied in as instances. The instructional materials would then include the rules for doing the task as well as an adequate number of examples (see Wandersee, Chap. 2 in this book).

Empirical Support

We have begun to evaluate the effectiveness of conceptual graph analysis for instructional design. In a recent study, we used three of the four knowledge acquisition methods to develop instructional materials in the area of engineering mechanics; document analysis, question probes, and inductive analysis [5]. Our primary research question was: Will the use of conceptual graph analysis result in greater learning than the instructional design method currently used in almost all educational and training programs, namely instructional materials written by an expert in the field and edited by peers?

The experimental design was therefore a simple comparison of two instructional systems; (1) text and graphics written by a recognized expert and edited by peers, and (2) text and graphics covering the same content but obtained by using conceptual graph analysis.

Using a popular text in engineering dynamics, we selected seven pages of relatively autonomous material on the topic of work-energy. Two designers independently graphed the text (with 81% agreement) and developed question probes to administer to the class instructor. The instructor answered the probes as well as more high-level questions such as "What is the relationship between this branch of information and the rest of the material?" and also solved several related domain problems. Based on the answers, the graph was pruned, expanded, and/or clarified. The graph was then rewritten as text using the same order of presentation as the original text.

When students were to study the work/energy section in class, they were given at random either the original text or the knowledge engineered text, both in the same format and font. After turning in the text, they solved four problems, three of which were used for the analysis and they also filled out an evaluation questionnaire. In this study, students in both groups took about the same amount of time to read the material. However, students who read the original text received an average of 2.9 points out of 9 possible, while students who read the knowledge engineered text received an average of 4 points. Looking at the number of problems solved by all students in each group, the twenty students in the original text group solved only 11 problems (out of 60), while students in the new text group solved 21 problems.

The conceptual graph analysis clearly improved the instructional text for the topic of work-energy. However, the problem solving scores were still lower than what an instructor would like to see. We feel that the scores most likely reflect either (a) the general difficulty of the topic, (b) the difficulties both groups of students experienced in translating 'conceptual' knowledge into problem solving knowledge, and/or (c) a lack of sufficient background knowledge for the material (the topic came late in the semester). Based on the evaluations, students liked the engineered text significantly better than the original expert-generated text. Both texts were described as not having enough examples, although the expert-generated text was rated more strongly so.

These data indicate that there are some definite advantages to using conceptual graph analysis to knowledge engineer instructional text, at least for some topics. However, in addition to increasing the completeness and coherence of the conceptual material (taxonomic, causal, etc.), a greater effort must be made to relate conceptual knowledge to problem solving goal structures. We are now using conceptual graph analysis to accomplish both of these goals for course textbooks in engineering mechanics, in particular in the areas of statics and dynamics.

Application to Teaching and Research in Biology

The methods described in this chapter can be applied to biology education in a number of ways. For the sake of discussion, the applications will be divided into several basic categories: use with students, instructional design, and research.

Use with Students

Conceptual graphs can be used in classroom or laboratory activities by individual students or by students working in groups. The graphs are used to make ideas explicit to students themselves as well as to the instructor. The conceptual graph structures can be drawn on paper or on computers using programs such as SemNet [6]. A special version of SemNet which allows propositions (long concept names) at nodes is required. Conceptual graph structures can be developed using free generation where students simply start writing or drawing. Alternatively, instructors can provide more specific question probes in verbal or written form. The use of question probes can be beneficial because it will usually elicit more of the student knowledge structures than free generation [4]. At times, it may be helpful for instructors to graph their own view of a particular topic and compare this with students.

Conceptual graphs are very detailed, and because of this they often make it apparent where students have conceptual difficulties. These are evidenced by one of several graph features such as missing nodes (such as important characteristics of a concept), missing relationships, and incorrect relationships. Incorrect relationships between nodes (node-arc-node triads) indicate misconceptions or other erroneous ideas. For example, in our VCR study, we found that almost all students misrepresented the TV/VCR button. They assumed that one has the button on 'TV' for watching TV, and on 'VCR' for watching tapes from the VCR. This was despite the fact that they had watched a very explicit and detailed instructional tape on the correct functioning and use of the button. This last example suggests the usefulness of using question probes and conceptual graphs to evaluate student knowledge before and after instructional activities.

Instructional Design

Conceptual graphs are a good tool for making knowledge explicit in instructional design. Both Brock Allen at San Diego State University and I have independently been teaching the use of conceptual graphs for concept and task analysis in instructional design [24]. The graphs are developed using the methods described earlier, and then translated into text, outlines, and other instructional forms such as hypertext. This approach has been successful in a variety of domains in both university and industry settings. Conceptual graphs can be developed on the basis of existing documents, the expertise of a single person, or used to integrate the expertise of several people. A major advantage of conceptual graph syntax is the ability to develop goal hierarchies for procedures to be taught.

Conceptual graph structures provide a visual diagram of the interrelationships (or lack thereof) between ideas. The fact that they have an inherent organization that reflects the underlying information base makes them especially appropriate for computer-based instructional systems such as hypertext. We have developed a cluster coherence algorithm to translate conceptual graphs into hypertext or hypermedia instructional systems. The algorithm specifies which graph nodes are to be grouped together into hypertext frames, and how frames are to be linked together [40, 41].

One might ask (and some have): "When is conceptual graph analysis most likely to be useful or necessary?" There are no cut and dried rules or answers to this question. However, it seems that graphical analysis would be most useful in the following circumstances:

(1) When students have historically shown difficulties in acquiring and/or using the information. One reason we have used the domain of VCRs for our research is the clear history of difficulties people have when using the more advanced VCR machines.
(2) The topic has a large number of abstract concepts and is being taught to novices.
(3) The information is complex. For example, in physics, the meaning and therefore understanding of a single concept, such as 'mass', depends on the interrelationship of several other abstract concepts.
(4) Performance of a task requires conceptual knowledge and has serious consequences resulting from error.

Conclusion

The methods described in this chapter can also be used to make knowledge structures explicit for educational research purposes. In evaluating learning, they can provide diagnostic tools to evaluate the development, organization, and use of student knowledge structures. They can also point out the differences between the novice, intermediate, and expert structures in terms of content and organization. However, it is clear that eliciting knowledge with question probes does not take the place of asking learners to actually use their knowledge in task performance.

It also does not take the place of modeling the dynamic cognitive processes in task performance and problem solving; as our research has indicated, students may have verbalizable knowledge and not be able to use it [2].

References

1. Gordon, S.E., Gill, R.T.: Question probes: A structured method for eliciting declarative knowledge. AI Appl in Nat Resource Mgt, 3, 13-20 (1989).
2. Gordon, S.E., Gill, R.T.: The formation and use of conceptual structures in problem-solving domains. Technical Report for AFOSR, grant #AFOSR-88-0063, 1989.
3. Gordon, S.E., Kinghorn, R., Schmierer, K.A.: Representing expert knowledge for instructional system design: A case study. Proc. of the Human Factors Soc., pp. 1412-1416. Santa Monica, CA: Human Factors Society 1991.
4. Gordon, S.E., Gill, R.T.: Knowledge acquisition with question probes and conceptual graph structures. In: Questions and information systems (T. Lauer, E. Peacock, A. Graesser, eds.), pp. 29-46. Hillsdale, NJ: Erlbaum 1992.
5. Gordon, S.E., Schmierer, K., Gill, R.T.: Conceptual graph analysis: Knowledge acquisition for instructional system design. Human Factors, in press.
6. Fisher, K.M.: Semantic networking: The new kid on the block. J. of Res. and Sci. Teaching, 27, 1001-1018 (1990).
7. Fisher, K.N., Fealetti, J., Patterson, H., Thornton, R., Lipson, J., Spring, C.: Computer-based concept mapping. J. of College Sci. Teaching, 19, 347-352, 1990.
8. Gowin, R., Novak, J. D.: Learning how to learn. New York: Cambridge University Press 1984.
9. Britton, B.K., Gulgoz, S.: Using Kintsch's computational model to improve instructional text: Effects of repairing inference calls on recall and cognitive structure. J. of Ed. Psych., 83, 329-345, 1991.
10. Goldberg, F., Bach, N.: A strategy for assessing the extent to which students can recall and apply scientific knowledge. Paper presented at Nat. Assoc. of Res. in Sci. Teaching, April 1991.
11. Kintsch, W., van Dijk, T.A.: Toward a model of text comprehension and production. Psych. Rev., 85, 363-394, 1978.
12. Graesser, A.C.: Prose comprehension beyond the word. New York: Springer-Verlag 1981.
13. Lenat, D.B., Guha, R.V., Pittman, K., Pratt, D., Shepherd, M.: Cyc: Toward programs with common sense. Communications of the ACM, 33, 30-49. New York: Association for Computing Machinery 1990.
14. Graesser, A.C., Clark, L.: Structures and procedures of implicit knowledge. Norwood, NJ: Ablex 1985.
15. Anderson, J.R.: Cognitive psychology and its implications. 3rd ed. New York: W.H. Freeman and Co. 1990.
16. Kintsch, W.: The role of knowledge in discourse comprehension: A construction-integration model. Psych. Rev., 95, 163-182, 1988.
17. Trabasso, T., Secco, T., Broek, P. V. D.: Causal cohesion and story coherence. In: Learning and comprehension of text (H. Mandl. N.L. Stein, T. Trabasso, eds.), pp. 83-111. Hillsdale, NJ: Erlbaum 1984.

18. Golding, J.M., Graesser, A.C., Millis, K.K.: What makes a good answer to a question?: Testing a psychological model of question answering. Discourse Processes, 13, 305-325, 1990.
19. Graesser, A.C., Franklin, S.P.: QUEST: A cognitive model of question answering. Discourse Processes, 13, 279-304, 1990.
20. Graesser, A.C., Goodman, S.M.: Implicit knowledge, question answering, and the representation of expository text. In: Understanding expository text (B. Britton, J.B. Black, eds.). Hillsdale, NJ: Erlbaum 1985.
21. Graesser, A.C., Goodman, S.M.: Research handbook: How to construct conceptual graph structures. In: Understanding expository text (B. Britton, J. B. Black, eds.). Hillsdale, NJ: Erlbaum 1985.
22. Graesser, A.C., Gordon, S.E.: Question answering and the organization of world knowledge. In: Memories, thoughts, and emotions: Essays in honor of George Mandler (G. Craik, A. Ortony, W. Kessen, eds.), pp 227-243. Hillsdale, NJ: Erlbaum 1991.
23. Graesser, A.C., Hemphill, D.H., Brainerd, L.E.: Question answering in the context of causal mechanisms. Proc. of the Cog. Sci. Soc., 621-626, 1989.
24. Moore, J., Gordon, S.E.: Conceptual graphs as instructional tools. Proc. of the Human Factors Soc., Vol. 2, pp 1289-1293. Santa Monica, CA: Human Factors Society 1988.
25. Graesser, A.C., Gordon, S.E., Sawyer, J.D.: Recognition memory for typical and atypical actions in scripted activities: Tests of a script pointer + tag hypothesis. J. of Verb. Learn. and Verb. Beh., 18, 319-332, 1979.
26. Roenker, D.L., Thompson, C. P., Brown, S. C.: Comparison of measures for the estimation of clustering in free recall. Psych. Bull., 76, 45-48, 1971.
27. Anderson, J.R.: The architecture of cognition. Cambridge, MA: Harvard University Press 1983.
28. Squire, L.M.: Memory and brain. New York, NY: Oxford University Press 1987.
29. Gordon, S.E.: Implications of cognitive theory for knowledge acquisition. In: The pychology of expertise: Cognitive research and empirical AI (R. Hoffman, ed.), pp. 99-120. New York: Springer-Verlag 1992.
30. Broadbent, D.E., Fitzgerald, P. Broadbent, M.H.P.: Implicit and explicit knowledge in the control of complex systems. Brit. J. of Psych., 77, 33-50, 1986.
31. Gagne, E.D.: The cognitive psychology of school learning. Boston, MA: Little, Brown and Co. 1985.
32. Schacter, D.L.: On the relation between memory and consciousness: Dissociable interactions and conscious experience. In: Varieties of memory and consciousness: Essays in honour of Endel Tulving (H.L. Roediger, F.I.M. Craik, eds.), pp. 355-389. Hillsdale, NJ: Erlbaum 1989.
33. Schneider, W., Detweiler, M.: Connectionist/control architecture for working memory. In: The psychology of learning and motivation (G.H. Bower, ed.). San Diego, CA: Academic Press 1987.
34. Schneider, W., Shiffrin, R.M.: Controlled and automatic human information processing: I. Detection, search, attention. Psych. Rev., 84, 1-66, 1977.
35. Carr, T.H., McCauley, C., Sperber, R.D., Parmalee, C.M.: Words, pictures, and priming: On semantic activation, conscious identification, and the automaticity of information processing. J. of Exp. Psych.: Hu. Perc. and Perf., 8, 757-777, 1982.
36. Marcel, A.T.: Conscious and unconscious perception: An approach to the relations between phenomenal experience and perceptual processes. Cog. Psych., 15, 238-300, 1983.

37. Means, B., Gott, S.P.: Cognitive task analysis as a basis for tutor development: Articulating abstract knowledge representations. In: Intelligent tutoring systems: Lessons learned (J. Psotka, L.D. Massey, S.A. Mutter, eds.). Hillsdale, NJ: Erlbaum 1988.
38. Sanderson, P.M.: Verbalizable knowledge and skilled task performance: Association, dissociation, and mental models. J. of Exp. Psych.: Learn., Mem., and Cog., 15, 729-747, 1989.
39. Lang, K.L., Graesser, A.C., Hemphill, D.D.: The role of questioning in knowledge engineering and the interface of expert systems. Poetics, 19 (1-2), 143-166, 1990.
40. Gordon, S.E., Lewis, V.: Knowledge engineering for hypertext instructional systems. Proc. of the Human Factors Soc., pp 1412-1416. Santa Monica, CA: Human Factors Society 1990.
41. Gordon, S.E., Gill, R.T.: Knowledge engineering for hypertext documents (unpublished paper under review).

Biological Models: Some Significant Features

John N. Honey

Centre for Educational Studies, King's College, University of London
Cornwall House Annex, Waterloo Road, London SE1 8TX, U.K.

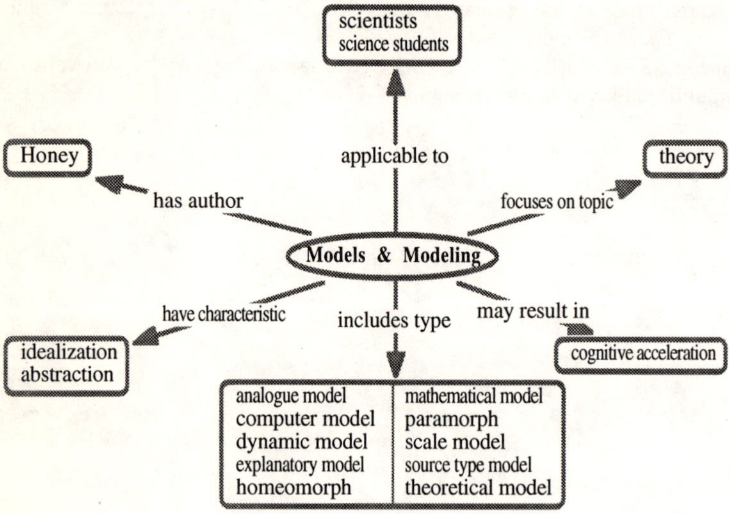

Abstract. The limitations of some different types of models are examined: in particular the relationships between a model and its subject. The importance of models and modeling in teaching and learning are discussed.

Keywords. Models, modeling

Some Features of Models

The importance of the construction and the use of models is continually emphasized in studies of science [8,11,20], the teaching and learning of science [7,18] and cognitive learning and understanding in general [17], although problems and difficulties which are specific to biology are not often mentioned. The purpose of this chapter is to identify some of the features of models and modeling which aid or hinder learning, and in particular those which are specific to biology. For this purpose, a number of features of biological models are examined and some characteristics of the structure of biological science are also considered. Harré defines a model as:

> some real or imagined thing or process which behaves similarly to some other thing or process, or in some other way than its behaviour is similar to it. [11]

This definition considers the model in relation to its subject but it may also be considered as a representation of someone's thought about the subject: a 'mental model'.

The thought or 'mental model' is represented by some verbal, symbolic or physical construction that can be seen or understood, which is used to aid in the comprehension of a subject which may not be seen or so well understood. Manipulation of parts of the model may be analogous to changes in aspects of the subject which it represents and may enable the user to make predictions about the subject. There are features of the model which are similar to those of the original subject: such as structures in a scale model or relationships in an analogue model. The ability to distinguish in what respect and to what degree some features are similar and others are not is of major importance in learning through models, and in using models. Models can be compared with other forms by which thought can be represented. There are a number of types of models. Although they do not necessarily mirror every detail of thought on a topic, they can be modified to reflect changed ideas and can give rise to imaginative predictions. Modeling is used in science as well as in science education and has been shown to contribute to the process of cognitive acceleration in science education [1].

Limitations of Different Types of Models

The similarities or significant differences between model and subject depend upon the type of model. The terms 'model' and 'modeling' are sometimes confusing. Some meanings are shared, but there are also instances where scientists, science educators and cognitive scientists either use the terms in different ways to meet different needs or they refer to different types of models. Models have been classified in various ways by Black [4] and Harré [10,11], according to their source (the same as, or different from the subject) type, (scale model, analogue model or theoretical model) and the elements by which models are represented (a solid, words, sentences, mathematical symbols, computer symbols and other symbols).

Harré uses the term paramorph to define models in which the subject is not the source of the model. These are frequently constructed with the aim of modeling processes, such as the pump model of ion transfer across a membrane and electrical conduction as a model of nervous conduction. A possible disadvantage of the use of a paramorph is that details of the model which are not strictly analogous with features of the subject are used in the construction of theories.

Scale models can reflect known proportional relationships in the subject and proportional differences about external factors are important in some, such as a solid model of a moving fish but is unimportant in, say, an enlarged solid model illustrating the shape of a red blood cell. The physical structure at times is a symbolic representation of a theoretical model and the three-dimensional structure is, therefore, a model of a model. The theoretical model must be understood by the learner in order to comprehend the solid model. The use of solid does not always bring about expected gains in learning, however. In one study models Seddon and Moore [24] found that the use of solid molecular models did not enhance learning. In this example the students who did not use models learned the topic more rapidly than those who did.

Particular features are emphasized in some groups of models. They include idealizations which may have the same series of properties as those of the subject, but those of the model are more perfect than those of the subject on some accepted scale of values. The model of simple genetic recombination which did not include mutation or linkage could be an example. Abstractions are models in which only some of the properties of the subject are selected: as in a coloured wire model of the blood system. Limitations are found in both idealizations or abstractions. The larger the number of variables that are excluded from the model, the greater the possibility of misunderstanding by the learner.

Mathematical modeling is important in science and to some extent in science education. Although most models contain some quantitative elements, in mathematical models an attempt is made to describe all features in mathematical terms. Features of mathematical modeling, which are common to other modeling processes, include the identification of variable factors and the framing of hypotheses. The formal representation of the subject can be achieved by the adoption of standard modeling procedures [9] which have applications, for example, in computer modeling. Mathematical models have the advantages of precision and ease of inference. However, they may have disadvantages in that selection of features of a biological system which may be described in mathematical symbols, and can impose constraints upon creative thinking about less quantitative relationships.

The use of a computer model in science education e.g., the Dynamic Modeling System [21], Cellular Modeling System [14], and STELLA [12], frequently provides opportunities for a student to model ideas or relationships in a formal manner using symbols, and to explore the possible consequences of changes in relationships by manipulation of the system which is provided. The processes can exhibit similarities to some stages in scientific modeling: allowing for experiment in the relationship of parts and the testing of predictions. However, there may be constraints in the degree of flexibility which is available to the student who is limited by the given general structure and parameters of the system. The computerized system is itself a model and its manipulation and limitations could suggest to the student that similar features always occur in scientific modeling.

Case Studies as a Characteristic Feature of Biology

The type of model in which the subject of the model is also its source has frequent application in biology. Harré calls these homeomorphs and are particularly important in modeling structures. Differences between biology and the physical sciences may be seen in some uses of models [15]. Although reductionists [e.g., 23] claim that there is no basic difference between the biological and physical sciences, others have attempted to argue for a fundamental distinctiveness. In this connection, Polanyi [19] emphasized the boundaries between organisms and their environments, boundaries which separate not only in detail but also in overall function. Ayala [3] stressed the significance of teleology in biology 'organisms are systems directively organized'. Rosenberg [22] suggested that a distinction lies in the development in the physical sciences of universal laws or principles which result in the production of a theoretical model which may be applied in several situations, which may be adapted and which may provide opportunity to make predictions. He considered that this is also true in biology when universal principles are considered: in evolution or in some phenomenon which is described in terms of physical science, for example in molecular biology. In contrast, Rosenberg suggests that much of biology consists of case-studies or specific instances from which extensive generalizations cannot be made. The structure or functions of a narrowly defined group may then be used as a model of the features of a larger range of organisms, as in the use of specific examples or types. Individual organisms, processes or habitats are used as teleomorphic models. In biological 'case studies' the model, such as a type organism, is often used in an attempt to illustrate a generalization. Although, this can result in a student making inaccurate extrapolations. For example, if a sunflower stem or an earthworm are each used as examples of groups of organisms, this could lead a student to assume that all of their features are typical of angiosperms or annelids. An assumption that basic physiology of bacteria is always similar to that of *Escherichia coli* proved unfortunate. Other dangers of extrapolation from type organisms has been examined elsewhere [15].

However, the predominance of case-studies or specific instances is not peculiar to biology [8]. Giere carefully explains that theoretical models in physics are not only human constructions, but that they represent an idealized situation only conceived by approximation.

> Mechanics texts continually refer to such things as "the linear oscillator", "the free motion of a symmetrical rigid body", "the motion of a body subject only to a central gravitational force", and the like. Yet the texts themselves make clear that the paradigm examples of such systems fail to satisfy fully the

> equations by which they are described. No frictionless pendulum exists, nor does any body subject to no external forces whatsoever. [8]

In other words, the 'Universal Laws' with which Rosenberg characterizes the physical sciences are also generalizations about a range of individual cases. It is not only inaccurate for a biology student to extend the use of facts concerning a cockroach and apply them to all insects, it is also inaccurate to generalize from the formula for motion of a simple pendulum, which is assumed to be frictionless, with a very small swing and uniform gravitational force, to all real pendulums. Generalization from an example used as a model can lead to weaknesses in understanding any part of science, therefore, science teaching must include simultaneous emphasis upon easily generalized features in a model which is an exemplar and explanation of its unique features which cannot be applied elsewhere. The type organism in biology may carry all of these possible disadvantages.

The Link Between a Model and its Subject

A key feature in the use of models is an understanding of the relationships between the model and its subject. The pendulum example illustrates the view that even in mechanics the theoretical models may be idealizations which may differ in several ways from a real situation. Giere's view [8] is that the model and reality are both objects, one abstract and one real and that a statement of the relationship between them might be described as a theoretical hypothesis. This relationship would better be described in terms of similarity instead of isomorphic. Specifically, the relevant respects and degree of similarity have to be clearly specified. He also suggests that a theory might be viewed as a cluster or even a population of models, and that a hypothesis may be judged as more or less accurately distinguishing the similarities which exist between the model and reality. In this view, the significance of a model is not that it exactly represents thought in the theoretical model but that it is one expression of the theoretical understanding of the subject or reality.

If, for example, a gene is used as an example of a model and is considered in relation to four elements of a model described by Harré [10], then each of these aspects can be viewed as a theoretical hypothesis linking the model and its subject. He lists:

(a) the existence of an entity, e.g., that genes exist,
(b) some of the attributes of the entity: genes are transmitted from parents to offspring,
(c) statements which assert a causal connection between the hypothetical entity and observed phenomena: genes are responsible for the existence of aspects of the phenotype,
(d) arrangements occurring in the model reflect comparable configurations in the subject: symbolic representation of linkage and crossing over mirrors actual changes in genetic constitution of organisms.

The Role of Models in Teaching and Learning

It has been suggested, by Shayer and Adey [25], that the ability to make use of models is one skill which distinguishes a recognizable change in cognitive development between concrete, operational and formal operational stages of children's learning in science. If modeling skills depend upon a stage in development then this must be taken into account when learning is considered. In his more recent work Adey [1] gives modeling as one of the major activities leading to cognitive acceleration in science and includes metacognition consideration of the process of modeling, to be an important aspect of the activity. His results give support to the importance of modeling and thinking about modeling as valuable contributions to the process of learning science.

Novak [18] suggests that: *"we see science not as 'truth seeking' but rather as the construction of explanatory models that encompass increasingly wider ranges of phenomena, that is, models that are robust and parsimonious".* This links with constructivists views of learning. For example, Driver, Guesne and Tiberghien [6] state that:

> In teaching science we are leading pupils to 'see' phenomena and experimental situations in particular ways; to learn to wear scientists' 'conceptual spectacles'. This involves pupils in constructing mental models for entities which are not perceived directly, such as light, electric current, particles of matter. The modeling process which is involved is complex; it requires pupils to construct and use certain entities, which may be sets of objects or systems, to describe these in exact ways using certain parameters (e.g. mass, volume, temperature, charge) and to account for the process of interaction between the parameters by describing relationships between them (using ideas such as forces, heat, electric current).'

Driver and Oldham [7] describe the 'active construction of meaning' by a student and stated that

> The constructions are seen as tentative models which are continually tested against experience and if necessary modified.

A student's understanding of respiration as merely the exchange of gases in breathing would be an example. With developed understanding the model may come to include some specific attributes of the process, some statements which assert a causal connection between the hypothetical process and observed phenomena, and some ways in which changes in the model may reflect comparable changes in the subject. The model of respiration could develop to include an understanding of the breakdown of food substances to release energy. The relationships within the model might be expressed in the form of equations, and changes in the compounds represented in the equation could be linked with experimental results. An example of the integration of teaching the subject of photosynthesis and the process of modeling, together with opportunity for students to develop their own models, was produced in material for use in the National Curriculum in Science for England and Wales [16]. Learning science involves both restructuring personal models and the development of an ability to describe the respects and degree to which the model and reality are similar. It might be suggested that either of these processes could be more easily achieved if the student is in a position to formalize the elements within the model, as in some computer modeling systems. The degree to which the normalization of the model, either in words or symbols, is important to students at various stages of learning, may prove interesting as studies of student's use of computer tools in modeling have shown [5].

This normalization is a result of initial mental modeling. In his discussion of mental models as a part of common human thought processes Johnson-Laird [17] says:

> The theory of mental models is intended to explain the higher processes of cognition and, in particular, comprehension and inference. It suggests a simple three-part inventory for the contents of the mind: there are recursive procedures, propositional representations and models.

He takes the view that these attributes, rather than the processes of classical logic are used in reasoning. If modeling in science is a particular example of the mental processes which Johnson-Laird describes, then logical construction of a model using precisely defined formal structures may only give a partial reflection of the cognitive process of mental modeling.

In a discussion of students' attitudes towards the nature of science, Hodson [13] distinguishes between the realist view of a theory and an instrumentalist view of models:

> In naive realism scientific theory is believed to provide a true description of the world. For the instrumentalist the observable, real world is described by means of an imaginary scientific theoretical model.

He quotes the statement that:

> Use of the terms 'model' and 'theory' within the science curriculum should, therefore, be an indication of the 'degree of certainty' with which we hold a particular view [26]

and adds:

> Difference in status between theories and models is not one of degree. Both are interpretations of sense data (not direct perceptions, but theories are believed to be 'true' (in the Popperian sense of 'corresponding to the facts'), models are not. Use of the terms model and 'theory' should, therefore, be an indication of the 'degree of certainty' with which we would hold a particular view.

Readhead [20] also emphasized that the process of modeling by a scientist may include the extraction of features of a theory to form a model which may then be tested:

> Science depends on the possibility of ignoring accidents, of isolating certain key features in a situation. These are captured by model, although in the very

act of idealization and approximation we convince ourselves that the model is indeed false.

Modeling could lead to a more critical consideration of the nature of science on the part of students as well as making a positive contribution to the process of learning science. Modeling may also give insight into the process of learning science [3] and the processes of thought involved in the conduct of scientific enquiry.

Interesting topics remain for exploration such as: how does modeling by a scientist compare with that of the student? and, how does the description of modeling as a part of the structure of science relate to a description of modeling as a mental process?

References

1. Adey, P.: The CASE results. Implication for science teaching. Int. J. Sci. Educ., 14, 137-146, 1991.
2. Association of Science Education, ASE, Hatfield Herts, UK.
3. Ayala, F.J.: Biology as an autonomous science. In: Topics in the philosophy of biology. (M.Grene and E.Mendelshon, eds.). Boston studies in the philosophy of science, Vol. 27, pp. 312-329. Boston: D. Reidel. 1976.
4. Black, M.: Models and metaphors: Studies in language and philosophy. New York: Cornell University Press 1962.
5. Bliss, J.et al.: Reasoning supported by computational tools. Comp. Educ., 18, 1-9, 1992.
6. Driver, R., Guesne, E., Tiberghien, A.: Childrens' ideas in science. Milton Keynes: Open University Press 1985.
7. Driver, R., Oldham,V.: A constructivist approach to curriculum development in science. Stud. Sci. Educ., 13, 105-122, 1986.
8. Giere, R.N.: Explaining science: A cognitive approach. Chicago and London: The University of Chicago Press 1988.
9. Gold, H.J.: Mathematical modelling of biological systems. London: John Wiley 1977.
10. Harré, R.: The principles of scientific thinking. London: Macmillan 1970.
11. Harré, R.: The philosophies of science. Oxford: The University Press 1985.
12. Hartley, J.R.: Learning from computer based learning in science. Stud. Sci. Educ., 15, 55-76, 1988.
13. Hodson, D.: Philosophy of science, science, and science education. Stud. Sci. Educ., 12, 25-57, 1985.
14. Holland, D.: A software laboratory. Dynamical modelling and the cellular modelling system. Sch. Sci. Rev., 69. 407-416, 1988.
15. Honey, J.N.: Models in biology: Form and function. J. Biol. Educ., 22, 295-300, 1988.
16. Honey, J.N. (Ed.).: Investigating the nature of science. Harlow: Longman and Nuffield-Chelsea Curriculum Trust (1990).
17. Johnson-Laird, P.N.: Mental models: Towards a cognitive science of language, inference and consciousness. Cambridge, MA: Harvard University Press 1983.

18. Novak, J.D.: Learning science and the science of learning. Stud. Sci. Educ., 15, 77-101, 1988.
19. Polanyi, M.: Life's irreducable structure. In: Topics in the philosophy of biology. (M. Grene and E. Mendelshon, eds.). Boston studies in the philosophy of science, Vol. 27, pp. 128-154. Boston: D. Reidel. 1976.
20. Readhead, M.: Models in physics. Brit. J. Phil. Sci., 31, 145-163, 1980.
21. Robson, K., Wong, D.: Teaching and learning with the dynamical modelling system. Sch. Sci. Rev., 66, 682-695, 1985.
22. Rosenberg, A.: The structure of biological science. Cambridge: The University Press 1985.
23. Schaffner, K.F.: The Watson-Crick model and reductionism. In: Topics in the philosophy of biology. (M. Grene, E. Mendelshon, eds.). Boston studies in the philosophy of science, Vol. 27, pp. 101-127. Boston: D. Reidel. 1976.
24. Seddon, G.M., Moore, R.G.: An unexpected effect in the use of models for teaching visualisation of rotation in molecular structures. Eur. J. Sci. Ed., 8, 79-86, 1986.
25. Shayer, M., Adey, P.: Towards a science of science teaching. London: Heinemann 1981.
26. Tawney, D.A.: The nature of science and scientific inquiry. In: The art of the science teacher (C.R. Sutton, J.T. Haysom, eds.), pp. 19-29. London: McGraw-Hill 1974.

Retrospective Causal Reasoning (RCR) in Biology

Pinchas Tamir

School of Education and Israel Science Teaching Center
Hebrew University, Jerusalem, Israel

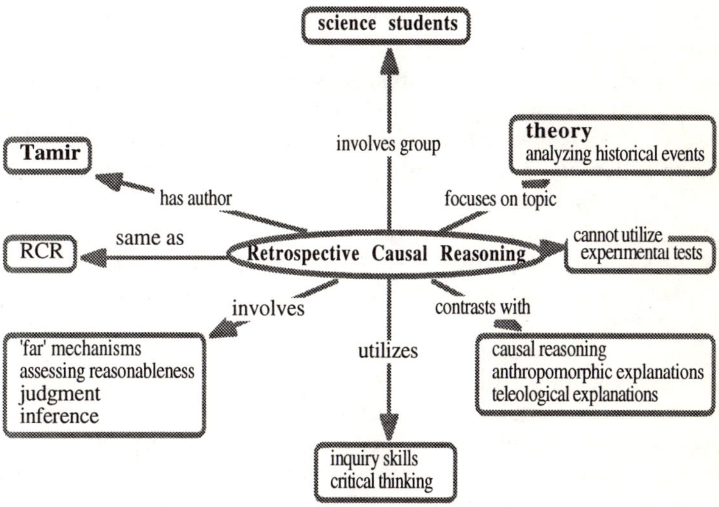

Abstract. The life sciences are defined and its nature and scope described in terms of unity-diversity, major sub-disciplines, main ideas and levels of organization. Special features of the life sciences are described under the four commonplaces, namely, subject matter (e.g., explanations, the world of micro); student (e.g., scientific literacy, alternative frameworks and conceptions); teachers (e.g., content specific pedagogy); and the milieu (e.g., outdoors, laboratory, Science Technology Society).

Keywords. Life sciences, unity, diversity, curriculum, structure and function, inquiry, conceptions, scientific literacy, alternative framework, attitudes, STS, intrinsic motivation, content specific pedagogy, problem solving

Introduction

Casual reasoning has been defined as "understanding why a cluster of properties have been grouped as an interactive unit and what processes maintain that interactive unit and thereby lend continuity to the kind" [3].

The purpose of causal reasoning is to identify the cause or causes of a particular event or phenomenon. By 'cause' we imply something that happened *prior* to the event, *before* the phenomenon has appeared. In living organisms the causes are typically biochemical or biophysical processes or mechanisms, which transfer from generation to generation by the DNA. Hence, when we wish to find out causes of events and phenomena, we are looking for such processes and mechanisms which existed in the organisms before the events take place. These mechanisms and processes are believed to have evolved through long periods of time. The ways in which such mechanisms have evolved are believed by most scientists to follow the principles of the theory of evolution. Since, according to this theory, there is always variation among individuals of the same population and only those individuals best adapted to their environment survive, many of the life processes and the responses of organisms appear as providing some benefits to the organisms. For example, when plants grow near a window they bend toward the window. By doing so they get more light which can be used in photosynthesis. Or another example, when trees shed their leaves in the fall and become dormant they are much more protected from damage of low temperatures than when they carry leaves and are not dormant. When asked for the causes for plants bending toward the window, many people will reply 'so that they can get more light for photosynthesis'. Similarly, when asked for the causes for trees to shed their leaves, many will reply 'so that they are protected from low temperature'. Such explanations which use the benefit as cause are designated as teleological. People who use them fail to distinguish between *cause* and *outcome*. They put the wagon before the horses without being aware of what they are doing.

There are two major reasons for the high frequency of teleological reasoning. The first is anthropomorphism or the tendency to view life processes as if they are governed by the same processes and mechanisms that control humans including consciousness and purposeful behavior. The second is the failure to distinguish between immediate and far mechanisms. In the case of the plants near the window, the 'immediate' mechanism involves different concentrations of growth hormones in the shaded and illuminated sides of the plant, whereas the 'far' mechanism is evolution. Only the first is to be referred to as causal, in the true sense of the term.

Some effective strategies which made students capable of distinguishing between causal, anthropomorphic and teleological explanations have been reported elsewhere [e.g., 2, 7].

One difference that has not been mentioned so far is that the causal explanation can be tested experimentally whereas the 'far' mechanism, even by virtue of the length of time involved alone, is not amenable to experimental studies but, rather, can only be inferred. The inference assumes that certain processes took place sometime in the past. Hence we call the kind of reasoning associated with inferred causes in the past, Retrospective Causal Reasoning (RCR).

Is RCR more difficult than simple causal reasoning? Does it involve unique cognitive processes? Is it the reason why many students have difficulty in comprehending and applying the theory of evolution? At present we have no answers to these questions. The purpose of this article is to draw attention to RCR and define some of its unique characteristics.

Where Else can RCR be Found?

Since we have not been able to locate studies on RCR in biological sciences, we looked for other fields of study in which RCR may be expected to take place. We examined causes related to accidents, law, crime, geology, archeology and history. The common feature for these fields of study is the need to collect evidence in the present about events that took place in the past with no way to test directly, such as by experiment, the validity of this evidence. This process is especially 'tricky' when cause and effect relations are to be established. In a very useful article on teaching history in high school, O'Reilly [4] suggests that students who seek to establish cause and effect in history should ask these two questions:

> (a) Did the historian show the connection between the proposed cause and effect, and (b) have other possible causes been eliminated? [4, p.372]

Whereas in history the evidence is in the form of documents and objects, in biology it is mostly based on observations. In both cases, however, there is a need to interpret and make inferences. These last processes are based on circumstantial evidence and cannot be used as hypotheses to be tested by experiments. Therefore the kind of information that is used as evidence is crucial. A good example of an activity designed to engage pupils in such reasoning is the story of the extinction of dinosaurs offered by Barman et al. [1]. Another example is a class investigation of a dead mouse presented by Shraer & Stalze [5].

Skills Involved in RCR

RCR involves the application of certain skills at assessing the reasonableness of ideas. According to Swartz [6] these skills may be classified in the following manner (Table 1).

Since no experiments can be carried out, what kind of prediction and generalization is intended? In the case of evolution an example of a hypothesis and prediction may be "if the rock is of a particular age, then fossils of particular organisms may be found in it," and a generalization would be "organisms that lived in water could use the dissolved oxygen in their respiration."

Table 1. Skills at assessing the reasonableness of ideas

(1) Support of Basic Information
- (A) Accurate Observation
- (B) Reliable/Unreliable Secondary Sources

(2) Inference
- (A) Use of Evidence
  - (1) Well/Ill-supported Explanation (Cause)
  - (2) Well/Ill-founded Prediction (Effect)
  - (3) Well/Ill-founded Generalization
- (B) Deductions
  - (1) Valid/Invalid Conditional Arguments
  - (2) Valid/Invalid Syllogistic Arguments

Thus it may be observed that retrospective causal reasoning involves all the processes of critical thinking and all the inquiry skills, with the exception of designing experiments to test hypotheses. Since experiments are the most powerful tests of hypotheses, the absence of experiments requires judgment which is an especially high level of reasoning when causes are inferred retrospectively. Moreover, whenever plausible, a particular problem should be approached from several different angles and evidence from different sources should be obtained, a process known as triangulation.

RCR and the Image of Science

Many textbooks present a scheme designated as the Scientific Method. Students are directed to believe as if only one well defined method exists. If one follows its steps in the right sequence, the desired outcomes will be obtained. Philosophers of science and science educàtors are interested in avoiding such a distorted image of science. Paying due attention to RCR, which is clearly incongruent with the classic 'Scientific Method' in which 'the crucial experiment' plays a decisive role, can help abolish the myth of the Scientific Method and demonstrate the legitimacy of plurality of methods in scientific research.

Possible Reasons for Difficulty of RCR

Students are taught to distinguish between correlation on the one hand and the cause/effect relationship on the other. Typically, the cause/effect relationship is established by a controlled experiment in which the independent variable is manipulated by the researcher. Under such controlled conditions the cause of the effect can be the only manipulated variable. The ability to control variables is an important educational objective to be developed during science studies. Once it has been mastered students know how to establish cause/effect relationships. This procedure is universally valid regardless of discipline and topics.

When studying past events, the standard method must be abandoned. Students are presented with a situation which cannot be handled by controlled experiment and they learn about cause/effect inferences which are based on correlational data. Whereas making inferences and conclusions based on experimental results is relatively simple and straightforward – involving either accepting or rejecting the tested hypotheses, presenting convincing evidence under RCR conditions may be very difficult and, quite often, controversial. Since the study of evolution involves RCR, it suffers from the difficulty just described. Yet, this is just one difficulty. Another difficulty, which is associated with the first one, is the supposed contradiction between the theory of evolution and the Biblical study of creation. People known as creationists use the fact that the powerful canon of evidence, the controlled experiment, is not available to scientists who study the theory of evolution, as an argument against the validity of the theory.

A third difficulty is the possible confusion between 'immediate' and 'far' causes described above in relation to teleological explanations. Students should understand the difference between causal and teleological explanations and why both are acceptable (e.g., Tamir & Zohar, 1991) before they can employ RCR effectively.

Teachers of biology and geology should be aware of the issues raised in this article in order to teach effectively topics which involve RCR.

References

1. Barman, C., DisSpezio, M., Guthrie, V., Leyden, M., Mercier, S., Ostlund, K.: Science. Menlo Park, CA: Addison Wesley 1989.
2. Bartov, H.: Can students be taught to distinguish between Teleological and causal explanations? Journal of Research in Science Teaching, 15, 567-572, 1978.
3. Keil, F.C.: Intuitive belief systems and informal reasoning in cognitive development. In: Information Reasoning and Education (J.F. Voss, D.N. Perkins & J.W. Segal, eds.) pp. 247-264. Hillsdale, NJ: Erlbaum 1990.
4. O'Reilly, K.: Informal reasoning in high school history. In: Information Reasoning and Education (J.F.Voss, D.N. Perkins, J.W. Segal, eds.), pp. 363-380. Hillsdale, NJ: Erlbaum 1990.
5. Schraer, R., & Stolze, J.: Biology: The study of life. Newton, MA: Allyn & Bacon 1987.
6. Swartz, R.J.: Structured teaching for critical thinking and reasoning in standard subject area instruction. In: Information Reasoning and Education (J.F. Voss, D.N. Perkins, J.W. Segal, eds.), pp. 415-450. Hillsdale, NJ: Erlbaum 1990.
7. Tamir, P., Zohar, A.: Anthropomorphism and teleology in reasoning about biological phenomena. Science Education, 75, 57-68, 1991.

Subject Index

The NATO ASI Series F Special Programme on ADVANCED EDUCATIONAL TECHNOLOGY

67: Designing Hypermedia for Learning
76: Multimedia Interface Design in Education
78: Integrating Advanced Technology into Technology Education
80: Intelligent Tutoring Systems for Foreign Language Learning
81: Cognitive Tools for Learning
84: Computer-Based Learning Environments and Problem Solving
85: Adaptive Learning Environments: Foundations and Frontiers
86: Intelligent Learning Environments and Knowledge Acquisition in Physics
87: Cognitive Modelling and Interactive Environments in Language Learning
89: Mathematical Problem Solving and New Information Technologies
90: Collaborative Learning Through Computer Conferencing
91: New Directions for Intelligent Tutoring Systems
92: Hypermedia Courseware: Structures of Communication and Intelligent Help
93: Interactive Multimedia Learning Environments
95: Comprehensive System Design: A New Educational Technology
96: New Directions in Educational Technology
97: Advanced Models of Cognition for Medical Training and Practice
104: Instructional Models in Computer-Based Learning Environments
105: Designing Environments for Constructive Learning
107: Advanced Educational Technology for Mathematics and Science
109: Advanced Educational Technology in Technology Education
111: Cognitive Models and Intelligent Environments for Learning Programming
112: Item Banking: Interactive Testing and Self-Assessment
113: Interactive Learning Technology for the Deaf
115: Learning Electricity and Electronics with Advanced Educational Technology
116: Control Technology in Elementary Education
119: Automating Instructional Design, Development, and Delivery
121: Learning from Computers: Mathematics Education and Technology
122: Simulation-Based Experiential Learning
125: Student Modelling: The Key to Individualized Knowledge-Based Instruction
128: Computer Supported Collaborative Learning
129: Human-Machine Communication for Educational Systems Design
132: Design of Mathematical Modelling Courses for Engineering Education
133: Collaborative Dialogue Technologies in Distance Learning
135: Technology Education in School and Industry
137: Technology-Based Learning Environments
138: Exploiting Mental Imagery with Computers in Mathematics Education
140: Automating Instructional Design: Computer-Based Development and Delivery Tools
141: Organizational Learning and Technological Change
142: Dialogue and Instruction. Modeling Interaction in Intelligent Tutoring Systems
146: Computers and Exploratory Learning
148: Knowledge Acquisition, Organization, and Use in Biology

NATO ASI Series F

Including Special Programmes on Sensory Systems for Robotic Control (ROB) and on Advanced Educational Technology (AET)

Vol. 25: Pyramidal Systems for Computer Vision. Edited by V. Cantoni and S. Levialdi. VIII, 392 pages. 1986. *(ROB)*

Vol. 26: Modelling and Analysis in Arms Control. Edited by R. Avenhaus, R. K. Huber and J. D. Kettelle. VIII, 488 pages. 1986. *(out of print)*

Vol. 27: Computer Aided Optimal Design: Structural and Mechanical Systems. Edited by C. A. Mota Soares. XIII, 1029 pages. 1987.

Vol. 28: Distributed Operating Systems. Theory und Practice. Edited by Y. Paker, J.-P. Banatre and M. Bozyiğit. X, 379 pages. 1987.

Vol. 29: Languages for Sensor-Based Control in Robotics. Edited by U. Rembold and K. Hörmann. IX, 625 pages. 1987. *(ROB)*

Vol. 30: Pattern Recognition Theory and Applications. Edited by P. A. Devijver and J. Kittler. XI, 543 pages. 1987.

Vol. 31: Decision Support Systems: Theory and Application. Edited by C. W. Holsapple and A. B. Whinston. X, 500 pages. 1987.

Vol. 32: Information Systems: Failure Analysis. Edited by J. A. Wise and A. Debons. XV, 338 pages. 1987.

Vol. 33: Machine Intelligence and Knowledge Engineering for Robotic Applications. Edited by A. K. C. Wong and A. Pugh. XIV, 486 pages. 1987. *(ROB)*

Vol. 34: Modelling, Robustness and Sensitivity Reduction in Control Systems. Edited by R.F. Curtain. IX, 492 pages. 1987.

Vol. 35: Expert Judgment and Expert Systems. Edited by J. L. Mumpower, L. D. Phillips, O. Renn and V. R. R. Uppuluri. VIII, 361 pages. 1987.

Vol. 36: Logic of Programming and Calculi of Discrete Design. Edited by M. Broy. VII, 415 pages. 1987.

Vol. 37: Dynamics of Infinite Dimensional Systems. Edited by S.-N. Chow and J. K. Hale. IX. 514 pages. 1987.

Vol. 38: Flow Control of Congested Networks. Edited by A. R. Odoni, L. Bianco and G. Szegö. XII, 355 pages. 1987.

Vol. 39: Mathematics and Computer Science in Medical Imaging. Edited by M. A. Viergever and A. Todd-Pokropek. VIII, 546 pages. 1988.

Vol. 40: Theoretical Foundations of Computer Graphics and CAD. Edited by R. A. Earnshaw. XX, 1246 pages. 1988. *(out of print)*

Vol. 41: Neural Computers. Edited by R. Eckmiller and Ch. v. d. Malsburg. XIII, 566 pages. 1988. *Reprinted as Springer Study Edition 1989, 1990.*

Vol. 42: Real-Time Object Measurement and Classification. Edited by A. K. Jain. VIII, 407 pages. 1988. *(ROB)*

Vol. 43: Sensors and Sensory Systems for Advanced Robots. Edited by P. Dario. XI, 597 pages. 1988. *(ROB)*

Vol. 44: Signal Processing and Pattern Recognition in Nondestructive Evaluation of Materials. Edited by C. H. Chen. VIII, 344 pages. 1988. *(ROB)*

Vol. 45: Syntactic and Structural Pattern Recognition. Edited by G. Ferraté, T. Pavlidis, A. Sanfeliu and H. Bunke. XVI, 467 pages. 1988. *(ROB)*

NATO ASI Series F

Including Special Programmes on Sensory Systems for Robotic Control (ROB) and on Advanced Educational Technology (AET)

Vol. 46: Recent Advances in Speech Understanding and Dialog Systems. Edited by H. Niemann, M. Lang and G. Sagerer. X, 521 pages. 1988.

Vol. 47: Advanced Computing Concepts and Techniques in Control Engineering. Edited by M. J. Denham and A. J. Laub. XI, 518 pages. 1988. *(out of print)*

Vol. 48: Mathematical Models for Decision Support. Edited by G. Mitra. IX, 762 pages. 1988.

Vol. 49: Computer Integrated Manufacturing. Edited by I. B. Turksen. VIII, 568 pages. 1988.

Vol. 50: CAD Based Programming for Sensory Robots. Edited by B. Ravani. IX, 565 pages. 1988. *(ROB)*

Vol. 51: Algorithms and Model Formulations in Mathematical Programming. Edited by S. W. Wallace. IX, 190 pages. 1989.

Vol. 52: Sensor Devices and Systems for Robotics. Edited by A. Casals. IX, 362 pages. 1989. *(ROB)*

Vol. 53: Advanced Information Technologies for Industrial Material Flow Systems. Edited by S. Y. Nof and C. L. Moodie. IX, 710 pages. 1989.

Vol. 54: A Reappraisal of the Efficiency of Financial Markets. Edited by R. M. C. Guimar ães, B. G. Kingsman and S. J. Taylor. X, 804 pages. 1989.

Vol. 55: Constructive Methods in Computing Science. Edited by M. Broy. VII, 478 pages. 1989.

Vol. 56: Multiple Criteria Decision Making and Risk Analysis Using Microcomputers. Edited by B. Karpak and S. Zionts. VII, 399 pages. 1989.

Vol. 57: Kinematics and Dynamic Issues in Sensor Based Control. Edited by G. E. Taylor. XI, 456 pages. 1990. *(ROB)*

Vol. 58: Highly Redundant Sensing in Robotic Systems. Edited by J. T. Tou and J. G. Balchen. X, 322 pages. 1990. *(ROB)*

Vol. 59: Superconducting Electronics. Edited by H. Weinstock and M. Nisenoff. X, 441 pages. 1989.

Vol. 60: 3D Imaging in Medicine. Algorithms, Systems, Applications. Edited by K. H. Höhne, H. Fuchs and S. M. Pizer. IX, 460 pages. 1990. *(out of print)*

Vol. 61: Knowledge, Data and Computer-Assisted Decisions. Edited by M. Schader and W. Gaul. VIII, 421 pages. 1990.

Vol. 62: Supercomputing. Edited by J. S. Kowalik. X, 425 pages. 1990.

Vol. 63: Traditional and Non-Traditional Robotic Sensors. Edited by T. C. Henderson. VIII, 468 pages. 1990. *(ROB)*

Vol. 64: Sensory Robotics for the Handling of Limp Materials. Edited by P. M. Taylor. IX, 343 pages. 1990. *(ROB)*

Vol. 65: Mapping and Spatial Modelling for Navigation. Edited by L. F. Pau. VIII, 357 pages. 1990. *(ROB)*

Vol. 66: Sensor-Based Robots: Algorithms and Architectures. Edited by C. S. G. Lee. X, 285 pages. 1991. *(ROB)*

Vol. 67: Designing Hypermedia for Learning. Edited by D. H. Jonassen and H. Mandl. XXV, 457 pages. 1990. *(AET)*

Vol. 68: Neurocomputing. Algorithms, Architectures and Applications. Edited by F. Fogelman Soulié and J. Hérault. XI, 455 pages. 1990.

Vol. 69: Real-Time Integration Methods for Mechanical System Simulation. Edited by E. J. Haug and R. C. Deyo. VIII, 352 pages. 1991.

NATO ASI Series F

Including Special Programmes on Sensory Systems for Robotic Control (ROB) and on Advanced Educational Technology (AET)

Vol. 70: Numerical Linear Algebra, Digital Signal Processing and Parallel Algorithms. Edited by G. H. Golub and P. Van Dooren. XIII, 729 pages. 1991.

Vol. 71: Expert Systems and Robotics. Edited by T. Jordanides and B.Torby. XII, 744 pages. 1991.

Vol. 72: High-Capacity Local and Metropolitan Area Networks. Architecture and Performance Issues. Edited by G. Pujolle. X, 536 pages. 1991.

Vol. 73: Automation and Systems Issues in Air Traffic Control. Edited by J. A. Wise, V. D. Hopkin and M. L. Smith. XIX, 594 pages. 1991.

Vol. 74: Picture Archiving and Communication Systems (PACS) in Medicine. Edited by H. K. Huang, O. Ratib, A. R. Bakker and G. Witte. XI, 438 pages. 1991.

Vol. 75: Speech Recognition and Understanding. Recent Advances, Trends and Applications. Edited by P. Laface and Renato De Mori. XI, 559 pages. 1991.

Vol. 76: Multimedia Interface Design in Education. Edited by A. D. N. Edwards and S. Holland. XIV, 216 pages. 1992. *(AET)*

Vol. 77: Computer Algorithms for Solving Linear Algebraic Equations. The State of the Art. Edited by E. Spedicato. VIII, 352 pages. 1991.

Vol. 78: Integrating Advanced Technology into Technology Education. Edited by M. Hacker, A. Gordon and M. de Vries. VIII, 185 pages. 1991. *(AET)*

Vol. 79: Logic, Algebra, and Computation. Edited by F. L. Bauer. VII, 485 pages. 1991.

Vol. 80: Intelligent Tutoring Systems for Foreign Language Learning. Edited by M. L. Swartz and M. Yazdani. IX, 347 pages. 1992. *(AET)*

Vol. 81: Cognitive Tools for Learning. Edited by P. A. M. Kommers, D. H. Jonassen, and J. T. Mayes. X, 278 pages. 1992. *(AET)*

Vol. 82: Combinatorial Optimization. New Frontiers in Theory and Practice. Edited by M. Akgül, H. W. Hamacher, and S. Tüfekçi. XI, 334 pages. 1992.

Vol. 83: Active Perception and Robot Vision. Edited by A. K. Sood and H. Wechsler. IX, 756 pages. 1992.

Vol. 84: Computer-Based Learning Environments and Problem Solving. Edited by E. De Corte, M. C. Linn, H. Mandl, and L. Verschaffel. XVI, 488 pages. 1992. *(AET)*

Vol. 85: Adaptive Learning Environments. Foundations and Frontiers. Edited by M. Jones and P. H. Winne. VIII, 408 pages. 1992. *(AET)*

Vol. 86: Intelligent Learning Environments and Knowledge Acquisition in Physics. Edited by A. Tiberghien and H. Mandl. VIII, 285 pages. 1992. *(AET)*

Vol. 87: Cognitive Modelling and Interactive Environments. With demo diskettes (Apple and IBM compatible). Edited by F. L. Engel, D. G. Bouwhuis, T. Bösser, and G. d'Ydewalle. IX, 311 pages. 1992. *(AET)*

Vol. 88: Programming and Mathematical Method. Edited by M. Broy. VIII, 428 pages. 1992.

Vol. 89: Mathematical Problem Solving and New Information Technologies. Edited by J. P. Ponte, J. F. Matos, J. M. Matos, and D. Fernandes. XV, 346 pages. 1992. *(AET)*

Vol. 90: Collaborative Learning Through Computer Conferencing. Edited by A. R. Kaye. X, 260 pages. 1992. *(AET)*

Vol. 91: New Directions for Intelligent Tutoring Systems. Edited by E. Costa. X, 296 pages. 1992. *(AET)*

NATO ASI Series F

Including Special Programmes on Sensory Systems for Robotic Control (ROB) and on Advanced Educational Technology (AET)

Vol. 92: Hypermedia Courseware: Structures of Communication and Intelligent Help. Edited by A. Oliveira. X, 241 pages. 1992. *(AET)*

Vol. 93: Interactive Multimedia Learning Environments. Human Factors and Technical Considerations on Design Issues. Edited by M. Giardina. VIII, 254 pages. 1992. *(AET)*

Vol. 94: Logic and Algebra of Specification. Edited by F. L. Bauer, W. Brauer, and H. Schwichtenberg. VII, 442 pages. 1993.

Vol. 95: Comprehensive Systems Design: A New Educational Technology. Edited by C. M. Reigeluth, B. H. Banathy, and J. R. Olson. IX, 437 pages. 1993. *(AET)*

Vol. 96: New Directions in Educational Technology. Edited by E. Scanlon and T. O'Shea. VIII, 251 pages. 1992. *(AET)*

Vol. 97: Advanced Models of Cognition for Medical Training and Practice. Edited by D. A. Evans and V. L. Patel. XI, 372 pages. 1992. *(AET)*

Vol. 98: Medical Images: Formation, Handling and Evaluation. Edited by A. E. Todd-Pokropek and M. A. Viergever. IX, 700 pages. 1992.

Vol. 99: Multisensor Fusion for Computer Vision. Edited by J. K. Aggarwal. XI, 456 pages. 1993. *(ROB)*

Vol. 100: Communication from an Artificial Intelligence Perspective. Theoretical and Applied Issues. Edited by A. Ortony, J. Slack and O. Stock. XII, 260 pages. 1992.

Vol. 101: Recent Developments in Decision Support Systems. Edited by C. W. Holsapple and A. B. Whinston. XI, 618 pages. 1993.

Vol. 102: Robots and Biological Systems: Towards a New Bionics? Edited by P. Dario, G. Sandini and P. Aebischer. XII, 786 pages. 1993.

Vol. 103: Parallel Computing on Distributed Memory Multiprocessors. Edited by F. Özgüner and F. Erçal. VIII, 332 pages. 1993.

Vol. 104: Instructional Models in Computer-Based Learning Environments. Edited by S. Dijkstra, H. P. M. Krammer and J. J. G. van Merriënboer. X, 510 pages. 1993. *(AET)*

Vol. 105: Designing Environments for Constructive Learning. Edited by T. M. Duffy, J. Lowyck and D. H. Jonassen. VIII, 374 pages. 1993. *(AET)*

Vol. 106: Software for Parallel Computation. Edited by J. S. Kowalik and L. Grandinetti. IX, 363 pages. 1993.

Vol. 107: Advanced Educational Technologies for Mathematics and Science. Edited by D. L. Ferguson. XII, 749 pages. 1993. *(AET)*

Vol. 108: Concurrent Engineering: Tools and Technologies for Mechanical System Design. Edited by E. J. Haug. XIII, 998 pages. 1993.

Vol. 109: Advanced Educational Technology in Technology Education. Edited by A. Gordon, M. Hacker and M. de Vries. VIII, 253 pages. 1993. *(AET)*

Vol. 110: Verification and Validation of Complex Systems: Human Factors Issues. Edited by J. A. Wise, V. D. Hopkin and P. Stager. XIII, 704 pages. 1993.

Vol. 111: Cognitive Models and Intelligent Environments for Learning Programming. Edited by E. Lemut, B. du Boulay and G. Dettori. VIII, 305 pages. 1993. *(AET)*

Vol. 112: Item Banking: Interactive Testing and Self-Assessment. Edited by D. A. Leclercq and J. E. Bruno. VIII, 261 pages. 1993. *(AET)*

Vol. 113: Interactive Learning Technology for the Deaf. Edited by B. A. G. Elsendoorn and F. Coninx. XIII, 285 pages. 1993. *(AET)*

NATO ASI Series F

Including Special Programmes on Sensory Systems for Robotic Control (ROB) and on Advanced Educational Technology (AET)

Vol. 114: Intelligent Systems: Safety, Reliability and Maintainability Issues. Edited by O. Kaynak, G. Honderd and E. Grant. XI, 340 pages. 1993.

Vol. 115: Learning Electricity and Electronics with Advanced Educational Technology. Edited by M. Caillot. VII, 329 pages. 1993. *(AET)*

Vol. 116: Control Technology in Elementary Education. Edited by B. Denis. IX, 311 pages. 1993. *(AET)*

Vol. 117: Intelligent Learning Environments: The Case of Geometry. Edited by J.-M. Laborde. VIII, 267 pages. 1996. *(AET)*

Vol. 118: Program Design Calculi. Edited by M. Broy. VIII, 409 pages. 1993.

Vol. 119: Automating Instructional Design, Development, and Delivery. Edited by. R. D. Tennyson. VIII, 266 pages. 1994. *(AET)*

Vol. 120: Reliability and Safety Assessment of Dynamic Process Systems. Edited by T. Aldemir, N. O. Siu, A. Mosleh, P. C. Cacciabue and B. G. Göktepe. X, 242 pages. 1994.

Vol. 121: Learning from Computers: Mathematics Education and Technology. Edited by C. Keitel and K. Ruthven. XIII, 332 pages. 1993. *(AET)*

Vol. 122: Simulation-Based Experiential Learning. Edited by D. M. Towne, T. de Jong and H. Spada. XIV, 274 pages. 1993. *(AET)*

Vol. 123: User-Centred Requirements for Software Engineering Environments. Edited by D. J. Gilmore, R. L. Winder and F. Détienne. VII, 377 pages. 1994.

Vol. 124: Fundamentals in Handwriting Recognition. Edited by S. Impedovo. IX, 496 pages. 1994.

Vol. 125: Student Modelling: The Key to Individualized Knowledge-Based Instruction. Edited by J. E. Greer and G. I. McCalla. X, 383 pages. 1994. *(AET)*

Vol. 126: Shape in Picture. Mathematical Description of Shape in Grey-level Images. Edited by Y.-L. O, A. Toet, D. Foster, H. J. A. M. Heijmans and P. Meer. XI, 676 pages. 1994.

Vol. 127: Real Time Computing. Edited by W. A. Halang and A. D. Stoyenko. XXII, 762 pages. 1994.

Vol. 128: Computer Supported Collaborative Learning. Edited by C. O'Malley. X, 303 pages. 1994. *(AET)*

Vol. 129: Human-Machine Communication for Educational Systems Design. Edited by M. D. Brouwer-Janse and T. L. Harrington. X, 342 pages. 1994. *(AET)*

Vol. 130: Advances in Object-Oriented Database Systems. Edited by A. Dogac, M. T. Özsu, A. Biliris and T. Sellis. XI, 515 pages. 1994.

Vol. 131: Constraint Programming. Edited by B. Mayoh, E. Tyugu and J. Penjam. VII, 452 pages. 1994.

Vol. 132: Mathematical Modelling Courses for Engineering Education. Edited by Y. Ersoy and A. O. Moscardini. X, 246 pages. 1994. *(AET)*

Vol. 133: Collaborative Dialogue Technologies in Distance Learning. Edited by M. F. Verdejo and S. A. Cerri. XIV, 296 pages. 1994. *(AET)*

Vol. 134: Computer Integrated Production Systems and Organizations. The Human-Centred Approach. Edited by F. Schmid, S. Evans, A. W. S. Ainger and R. J. Grieve. X, 347 pages. 1994.

Vol. 135: Technology Education in School and Industry. Emerging Didactics for Human Resource Development. Edited by D. Blandow and M. J. Dyrenfurth. XI, 367 pages. 1994. *(AET)*

Vol. 136: From Statistics to Neural Networks. Theory and Pattern Recognition Applications. Edited by V. Cherkassky, J. H. Friedman and H. Wechsler. XII, 394 pages. 1994.

NATO ASI Series F

Including Special Programmes on Sensory Systems for Robotic Control (ROB) and on Advanced Educational Technology (AET)

Vol. 137: Technology-Based Learning Environments. Psychological and Educational Foundations. Edited by S. Vosniadou, E. De Corte and H. Mandl. X, 302 pages. 1994. *(AET)*

Vol. 138: Exploiting Mental Imagery with Computers in Mathematics Education. Edited by R. Sutherland and J. Mason. VIII, 326 pages. 1995. *(AET)*

Vol. 139: Proof and Computation. Edited by H. Schwichtenberg. VII, 470 pages. 1995.

Vol. 140: Automating Instructional Design: Computer-Based Development and Delivery Tools. Edited by R. D. Tennyson and A. E. Barron. IX, 618 pages. 1995. *(AET)*

Vol. 141: Organizational Learning and Technological Change. Edited by C. Zucchermaglio, S. Bagnara and S. U. Stucky. X, 368 pages. 1995. *(AET)*

Vol. 142: Dialogue and Instruction. Modeling Interaction in Intelligent Tutoring Systems. Edited by R.-J. Beun, M. Baker and M. Reiner. IX, 368 pages. 1995. *(AET)*

Vol. 144: The Biology and Technology of Intelligent Autonomous Agents. Edited by Luc Steels. VIII, 517 pages. 1995.

Vol. 146: Computers and Exploratory Learning. Edited by A. A. diSessa, C. Hoyles and R. Noss. VIII, 482 pages. 1995. *(AET)*

Vol. 147: Speech Recognition and Coding. New Advances and Trends. Edited by Antonio J. Rubio Ayuso and J. M. López Soler. XI, 505 pages. 1995.

Vol. 148: Knowledge Acquisition, Organization, and Use in Biology. Edited by Kathleen M. Fisher and Michael R. Kibby. X, 246 pages. 1996. *(AET)*

Springer-Verlag and the Environment

We at Springer-Verlag firmly believe that an international science publisher has a special obligation to the environment, and our corporate policies consistently reflect this conviction.

We also expect our business partners – paper mills, printers, packaging manufacturers, etc. – to commit themselves to using environmentally friendly materials and production processes.

The paper in this book is made from low- or no-chlorine pulp and is acid free, in conformance with international standards for paper permanency.

Printing: Druckhaus Beltz, Hemsbach
Binding: Buchbinderei Schäffer, Grünstadt